T0212419

The Toxic Microbiome

Gut microbiomes are dynamic communities varying from population to population and throughout life. In Western societies, a toxic metabolic shift of gut microbiomes is a driver and underestimated risk factor for the development of many noncommunicable chronic pathologies. This book identifies the root cause of these deleterious microbial changes. During the last several decades, increased consumption of animal products, coinciding and correlated with global climate change, has been a contributing cause of undesirable gut microbiome changes.

Key Features

- Establishes a connection between poor gut microbiome health and chronic disease and cancer development
- Demonstrates how animal products and low-fiber diet patterns induce a detrimental metabolic transition of the gut microbiome from a human health-maintaining toward a disease-promoting state
- Discusses the opportunity of a toxic microbial metabolic signature as a powerful clinical and diagnostic tool to effectively predict chronic disease and cancer development
- Provides the latest evidence on different strategies to rebuild a healthy microbiome metabolism and effectively prevent noncommunicable diseases and colorectal cancer
- Documents the gut microbiome benefits of a plant-based diet

Related Titles

- Olds, W., ed. *Health and the Gut: The Emerging Role of Intestinal Microbiota in Disease and Therapeutics* (ISBN 978-1-7746-3204-8)
- Wilson, M. *The Human Microbiota in Health and Disease: An Ecological and Community-Based Approach* (ISBN 978-0-8153-4585-5)
- Schriffrin, E. J., et al., eds. *Intestinal Microbiota in Health and Disease: Modern Concepts* (ISBN 978-1-4822-2676-8)
- Lopata, A. L., ed. *Food Allergy: Molecular and Clinical Practice* (ISBN 978-0-3677-8199-6)

The Toxic Microbiome
Animal Products and the Demise of the Digestive Ecosystem

Sarah Schwitalla

CRC Press
Taylor & Francis Group
Boca Raton London New York

CRC Press is an imprint of the
Taylor & Francis Group, an **informa** business

First edition published 2023
by CRC Press
6000 Broken Sound Parkway NW, Suite 300, Boca Raton, FL 33487-2742

and by CRC Press
4 Park Square, Milton Park, Abingdon, Oxon, OX14 4RN

CRC Press is an imprint of Taylor & Francis Group, LLC
© 2023 Sarah Schwitalla

Library of Congress Cataloging-in-Publication Data

Names: Schwitalla, Sarah, author.
Title: The toxic microbiome : animal products and the demise of the digestive ecosystem / Sarah Schwitalla.
Description: First edition. | Boca Raton, FL : CRC Press, 2022. | Includes bibliographical references and index.
Identifiers: LCCN 2022018441 (print) | LCCN 2022018442 (ebook) | ISBN 9781032080000 (hbk) | ISBN 9781032065120 (pbk) | ISBN 9781003212447 (ebk)
Subjects: LCSH: Gastrointestinal system--Microbiology. | Gastrointestinal system--Diseases--Prevention. | Gastrointestinal system--Cancer--Prevention. | Digestive organs--Microbiology. | Vegetarianism. | Nutrition.
Classification: LCC QR171.G29 S39 2022 (print) | LCC QR171.G29 (ebook) | DDC 612.3/2--dc23/eng/20220817
LC record available at https://lccn.loc.gov/2022018441
LC ebook record available at https://lccn.loc.gov/2022018442

ISBN: 9781032080000 (hbk)
ISBN: 9781032065120 (pbk)
ISBN: 9781003212447 (ebk)

DOI 10.1201/9781003212447

Typeset in ITC New Baskerville Std
by KnowledgeWorks Global Ltd.

For
Victoria, Ladybird, Zlatan
and
LOML

CONTENTS

PREFACE

When I was 18, I decided I wanted to cure cancer.

More than 8 years into my journey of being a successful cancer and stem cell researcher, I had gained a pretty good understanding of how colorectal cancer cells form and how they metastasize. I had studied intra- and intercellular signaling, inflammatory processes in tumor tissues, I was able to grow tumor organoids in a petri dish and developed tumor mouse models, but I was unaware of the major risk factors that cause cancer. I realized, I was familiar with the tumor microenvironment, but I had no idea of the macro-environmental factors and how they actually initiate the cancerous process.

I knew *how* but not *why*.

I felt I was caught in a molecular ivory tower in the midst of a microscopic world of microbes, human cells, single molecules, genes and genetic fragments while being alienated from the real-world setting of cancer patients that I wanted to serve and of the people who accumulate environmental risk factors of which they were unaware of being identified long ago to be the actual villains responsible for more than half of all cancer cases in the population.

Genes load the gun, but lifestyle pulls the trigger, they say.

In 1971, President Richard Nixon declared the "war on cancer" dedicating a huge budget of millions of dollars in funding to heroically win the war. In 2019, research funds available to the National Cancer Institute of the NIH (National Institutes of Health) totaled $6.1 billion[1].

More than 50 years after Nixon's declaration, cancer incidence is
still high and rising and too many people are dying prematurely.
While it's clearly important to find out about the molecular
mechanisms of the disease to deliver the groundwork for patient
relevant care and drug-based treatment the question remains: How
can we prevent all that?

To win the war on cancer after all someday, to prevent people from
becoming patients and serve patients by improving their quality of
life better, we obviously need to know how exactly the gun works
but equally important what pulls its trigger.

Therefore, I wrote this book.

I aim to put molecular and biomedical research into a broader
public health context to facilitate interdisciplinary exchange across
different medical research and health care fields. Cancer is a
holistic disease that requires a strong interdisciplinary expertise in
order to offer patient-relevant care and efficiently benefit national
health care systems.

In my book, I try to provide some answers to the questions: What
are the root causes and what is the major risk factor for cancer and
chronic disease development, why and how do they exactly trigger
diseases? As the last decade of research revealed, the human gut
microbiome has increasingly been related to various chronic diseases
and cancer, however, little conclusive evidence for a causal role exists.

Therefore, this book also wishes to provide comprehensive
insights into the gut microbiome and its function in disease and
cancer development. Regarding the increasing complexity of
microbiome and cancer research, I intend to give an overview of
current evidence by extracting the most relevant resources out of

the tremendous paper flood of current literature. I will cover the risk factors that cause our global burden of disease, introduce and explore the major cellular players and their biochemistry involved in disease development and finally translate the findings into an optimistic outlook on public health and disease prevention.

I think this essential and important health issue should be accessible and available to everyone no matter which background. Based on scientific evidence, I want this book to be an engaging, informative and intriguing read without merely delivering a vast amount of defragmented information out of context in order to help the reader gain a comprehensive understanding of humanity's major burden of disease causes and mechanisms and offer evidence how to prevent it.

I would like to invite scientists, but also readers of all backgrounds, to keep reading and understand how our exposome and particularly diet (being the biggest risk factor of all) influences our body and our gut microbiome, how the microbiome can turn into a disease-promoting organ, and ultimately how to significantly lower the risk of developing chronic diseases and cancer.

AUTHOR

Sarah Schwitalla, PhD, is a passionate scientist dedicated to cancer prevention. She graduated from the Technical University Munich in Germany. An EMBO Scholar, she was a research fellow at Children's Hospital Boston and Harvard Medical School. She works as a consultant and founder at a digital cancer and gut disease prevention center (DACH region) with the mission to raise awareness about disease prevention in the population. She has published articles in *Cell, Cancer Cell, Nature, Cell Stem Cell* and *PNAS* and is the author of two books. She lives in Sweden with her husband.

INTRODUCTION

Today, 7 of 10 deaths will occur due to a noncommunicable disease[2].

Since the 1970s, because of the acceleration of globalization, industrialization and adaptation of a modern lifestyle, Western societies experienced a stark rise in chronic diseases and cancer incidence with emerging and developing countries already catching up. More than 70% of global deaths (32 million people) are attributable to cancers, cardiovascular diseases, chronic respiratory diseases and diabetes alone[3]. Cancer has become the second leading cause of death in the United States[4]. In fact, every second man and every third woman in the country will be diagnosed with cancer at some point during their lifetime and almost 90% of all men and women in Western societies nowadays have a higher probability of dying from a noncommunicable disease before the age of 70 than from communicable diseases or undernutrition[2,5].

In point of fact, noncommunicable diseases (NCDs) have become the leading cause of death and ill health worldwide[6].

Chronic diseases and cancer start early.

Years before noticeable symptoms appear, a disease has begun long ago on the molecular level. A still widely unexplored and underestimated key player in this drama turns out to be the human microbiome – the around 40 trillion microbes comprising nearly 60% of the cells in our human body[7]. Accumulating evidence suggests that the microbiome may be a mediator and influential factor in human chronic disease development. Although mostly correlative, there is accumulating evidence for certain microbial traits to be a dangerous trigger of cancer and disease development.

This is a book about finding answers. The central questions are: What is the major risk factor for developing cancer and chronic diseases? Which role does the human microbiome actually play? Is it a bystander of chronic disease and cancer development or is it even causally driving it? And if so, why and how?

In order to understand the relationship between the microbiome and NCD development, we need to get to know the major external risk factors and gain an in depth understanding of how they trigger the molecular mechanisms in the human holobiont that drive disease development.

This book aims to provide both.

In fact, microbiome health and public health are inevitably interconnected.

In Chapters 2, 5 and 6 we will uncover that the gut microbiome is critically sandwiched in between human exposure to risk factors and disease development: first, it becomes a victim itself before it turns into a toxic perpetrator contributing to disease progression and even its initiation. Chapters 1 and 5 demonstrate, that more than any environmental or genetic risk, an unhealthy diet pulls the trigger in this process. According to the Global Burden of Disease Study and the EAT Lancet commission unhealthy diets have consistently been found to "pose a greater risk to morbidity and mortality than does unsafe sex, and alcohol, drug, and tobacco use combined"[8]. Chapters 1, 5 and 6 describe how a skewed diet pattern consisting of a lack in fiber-rich plant-based foods crowded out by an abundance of animal products on our plates causes a microbiome dysbiosis, a metabolic shift from a health-maintaining towards a disease-promoting state.

This book also argues against the still prevailing concept of dysbiosis to be a shift in microbial species composition from "healthy" to "sick" but rather provides evidence for the arising toxic biochemical activities of the microbiota metabolism, that play a leading role in disease development as will be a topic of interest in Chapters 3 and 4.

We will also get to know the fundamentals of the microbiome and its functionality in human health in Chapter 4, which simultaneously proposes a paradigm shift in microbiome science. Chapter 3 describes how current microbiome science still focuses predominantly on certain taxa or a composition of species to be a marker of health or disease instead of its overall function. Unfortunately, this approach has led to broad inconsistencies and ambiguous results that despite the lack of clear evidence have even provoked some premature conclusions about a causal role of the microbiome in specific diseases. Consequently, Chapter 4 demonstrates that "who is there" may be not so important than what they do. A number of studies analyzed how individuals share only roughly 40% of the microbial species but more than 80% of microbial metabolic pathways and its metabolites, suggesting that "function rather than species" counts[9]. This view may offer an intriguing incentive for basic research as it can facilitate studying the microbiome in disease development but also accelerate translating the findings into clinical diagnostics. Monitoring microbial metabolites in the body rather than detecting which bacteria produce them may even become a valid, little invasive tool for effectively predicting disease development in the future as we will see in Chapter 6.

Chapter 8 includes some of the current research and clinical attempts to rebuild a healthy gut microbiome in patients, including probiotics and fecal microbial transplants.

Finally, the book also aims to provide answers to the most pressing question: How can we prevent the microbiome to become toxic in the first place and how can we prevent cancer and chronic disease development in the human body to lower the burden of disease in our societies? Chapter 7 proposes evidence-based, simple, safe and accessible measures that have consistently been demonstrated to be effective in reducing the individual risk of cancer and noncommunicable diseases and lower the burden of disease in populations.

We already have the most powerful tools at hand to save almost 11 million lives unnecessarily lost to largely preventable diseases each year, we just need to apply them.

Chapter

1

DIET-RELATED CHRONIC DISEASES ARE THE MOST CRITICAL HEALTH PROBLEM OF MODERN SOCIETIES

HOW DID THAT HAPPEN?

DOI: 10.1201/9781003212447-1

Eating a hot dog could cost you 36 minutes of healthy life; choosing to eat a serving of nuts instead could help you gain 26 minutes of extra life and cut your dietary carbon footprint by one-third[10].

The course of our life depends on the small decisions we make every day. Particularly, our daily food choices determine more than anything else whether we stay healthy or are prone to dying earlier from a chronic disease.

To date, diet-related chronic diseases, including obesity, type 2 diabetes, cardiovascular diseases, hypertension and stroke, chronic kidney diseases as well as various forms of cancers, are the leading causes of disability and premature death worldwide. Chronic noncommunicable diseases kill more than 41 million people around the globe each year, equivalent to 71% of all deaths[3]. Besides that, people spend many years of their life in a poor state of health. Too many happy life years are lost due to chronic noncommunicable diseases and cancer in Western countries such as the United States, the United Kingdom and Germany. According to the World Health Organization (WHO), 9 of the top 10 leading causes of disability and premature death are attributable to noncommunicable diseases in these countries[11]. Years of healthy life lost from diabetes solely increased by more than 80% between 2000 and 2019, while cardiovascular diseases rank on the first places at an all-time high in the United States and other Western countries. According to a comprehensive German analysis from 2021, the German population lost almost 12 million years of healthy life unnecessarily to diet-related disorders in 2017[12]. The US population even lost 60 million years in 2019[13]. Strikingly, nearly half of them already affect younger people under the age of 60.

Spending years of one's life in a chronic disease state became a global, rising trend in the last 50 years and turned out to be the "new normal". By 2025, it is projected that more than two-thirds of the years of life lost on our planet will be attributable to the consequences of poor diet and Western lifestyle[14]. According to the Global Burden of Disease Study around 11 million deaths, in the ballpark of populations in countries such as Sweden or Haiti, were attributable to poor diet in 2017[15].

WHEN AND WHY DID OUR LIFESTYLE AND OUR DIET BECOME SO DEADLY?

Since the 1960s, there is a heavy drift going on global plates. Being an aftermath of the World War II era, the food system and global dietary patterns have changed substantially, starting out in wealthier, Western countries based on an agenda of "productionism" and spreading to developing and emerging countries as a consequence of the "Green Revolution" agenda aiming to provide everyone with affordable, accessible food at all times[16]. As a major result of this food transformation, people began to center their diet around foods considered to be of "higher value", such as convenient, industrially processed and animal-based foods, while neglecting unprocessed plant-based options, which are associated with scarcity and frugality. According to the Food and Agricultural Organization of the United Nations (FAO), the consumption of calories from meat, sugar and processed fats (e.g., oils) increased globally by a multiple, while a dramatic decline in the consumption of pulses, root vegetables and tubers can be observed since the 1960s[17-20]. In contrast, total global meat consumption rose by around 60% since 1998, with a larger growth in developing countries than in Western counterparts due to their already high consumption[21,22].

Nowadays, the average American indulges in around 12 ounces of meat daily, but fails widely in meeting the minimum daily recommended amount of "5 a day" in fruits and vegetables[23,24].

Of note, diets low in fruit, high in sodium and low in whole grains together account for more than half of all diet-related deaths[15]. Likewise, the world eats only 12% of the daily recommended amount of nuts and seeds and only 23% of the recommended amount of whole grains per day, while people's intake of sugar-sweetened beverages and meat exceeds the recommended daily allowance by multiple times[15]. "This study affirms what many have thought for several years – that poor diet is responsible for more deaths than any other risk factor in the world", says Dr. Christopher Murray, Director of the Institute for Health Metrics and Evaluation at the University of Washington (USA), referring to recent data from the Global Burden of Disease Study. In fact, people's current diet priorities are a bigger contributor to mortality and morbidity than tobacco (6%) plus alcohol (5%) plus physical inactivity (3%)[14].

Although industrialization, the productionism paradigm and the "Green Revolution" of food systems increased the global food production in many countries and contributed to reducing global hunger, the type of foods produced did not necessarily contribute to our health, as rising numbers of chronically sick people tell. "The focus on increasing crop yields and improving production practices have contributed to reductions in hunger, improved life expectancy, falling infant and child mortality rates, and decreased global poverty. However, these health benefits are being offset by global shifts to unhealthy diets that are high in calories and heavily processed and animal source foods. These trends are driven partly by rapid urbanization, increasing incomes, and inadequate accessibility of nutritious foods", as a landmark paper of the Lancet EAT Commission explains the ongoing process of our global food transition within the recent decades[8].

Humans are literally eating themselves to death. Although more than 820 million people worldwide still have too little food and remain undernourished, being overfed and chronically sick turns out to be the world's most pressing health problem of all time[25]. More than 2 billion people are micronutrient deficient as a result poor diet quality, 2 billion adults are overweight or even obese and the global prevalence of type 2 diabetes almost doubled in the past 30 years[26-29]. Lifestyle-related chronic diseases took over long ago from more traditional public health concerns like undernutrition and infectious disease, and have placed an enormous burden on already overtaxed national health budgets in Western and emerging countries. In the United States alone, chronic diseases already consume 90% of annual health care expenditure[30].

Put even further into perspective, our poor food choices kill even more people around the world than any pathogenic microbe or infections. Around 2 million people die from tuberculosis, HIV/AIDS, malaria and cholera, primarily in developing countries; in contrast, 11 million pass away due to unfortunate food decisions in Western and emerging geographical regions[14,15,31].

According to the *Cancer Atlas*, cancer ranks either as the first or second leading cause of premature death in 134 countries of the world[4].

For years already, colon cancer turns out to be the third leading cause of cancer deaths in the United States and Europe. Around 150,000 Americans and 1.8 million people worldwide are affected[32]. Nearly one-third of them will die[33].

Modern screening methods have permitted early detection, luckily about less than 1% people are being diagnosed with a late stage, malignant colon cancer each year[32]. Unfortunately, the slight downward trend is mostly in older adults and masks the rising incidence among younger adults, which began in the mid-1990s.

Between 2012 and 2016, the incidence increased every year by 2% in people younger than 50 and 1% in people who are between 50 to 64[33].

Colon cancer is called a "Western disease", with an incidence in Americans of African descent of 65:100,000 compared to <10:100,000 in rural Africans[32,34]. Migrant studies, such as those on Japanese Hawaiians, have demonstrated that it only takes one generation for the immigrant population to reach the colon cancer incidence of the Western population[35].

Similarly, the emergence of chronic digestive disorders and inflammatory bowel diseases (IBDs) is quite a modern phenomenon and still affects mostly Western populations. Globally, IBDs concern more than 1 million people in the United States and around 2.5 million in Europe, resulting in significant health and economic costs[36]. Although the etiology of these diseases appear to be multifactorial, the corroborating evidence of the significant relationship between unhealthy dietary habits and corresponding lifestyle and IBD development is well acknowledged[37,38].

The majority of noncommunicable diseases and all cancer cases would be preventable through the reduction of the four main behavioral risk factors: smoking, physical inactivity, high alcohol intake and unhealthy dietary choices, according to the WHO and the World Cancer Research Fund (WCRF)[39-41]. A WCRF-initiated study published in 2021 found that 67% of colorectal cancer cases in the United Kingdom could have been preventable by these simple lifestyle adjustments[42]. The researchers figured the insufficient intake of dietary fiber as being the highest risk factor for colorectal cancer development, followed by meat consumption. Therefore, the WHO and a number of scientists consider primary prevention to be the most successful, cost-effective, accessible and sustainable course of action to tackle the chronic disease epidemic worldwide[39,40].

Despite the dramatic rise in the chronic disease burden and the clear awareness among public health authorities about the strong role of today's Western diet in disease development, apparently primary prevention is still underestimated and is not promoted well in the population.

Scientists and public health organizations such as the WHO and WCRF criticize the inaction by the countries and warn that rising chronic disease trends will inevitably result in a system failure and missing of the United Nations Sustainable Development Goals by 2030. After an assessment in 2021, the organizations came to the joint conclusion: "At present we are off-track to meet targets set out by the WHO for nutrition and NCDs by 2025, which means the Sustainable Development Goals are unlikely to be met by 2030, and sadly the situation has worsened since we have seen the two pandemics of NCDs and COVID-19 collide, making it extremely unlikely that targets will be met by all countries". At a 2021 WHO Executive Board meeting, they urged the international community to "accelerate efforts to develop and implement diet-related NCD (noncommunicable disease) policies, engage with civil society in order to help strengthen action networks" and "recognize and address actions by the food and beverage industry that undermine health"[43].

The global epidemic of diet-related chronic diseases turns out to be one of the most underrecognized public health crises to date. It is the major outcome of our skewed food choices, favoring processed and animal-based products over an unprocessed, plant-based diet.

We constantly deprive our body of vital nutrients needed for maintenance, a healthy metabolism and effective cell repair mechanisms, essential to protect us against diseases and cancer. However, it is not only our human cells that have become a victim of the evolution of unhealthy dietary choices since the 1960s. Likewise, the nutritional requirements for the trillions of microbial cells, comprising more than half of the cells in our body, are widely unmet. Since their most "favorite snack", plant fiber, is also their major energy source, the microbiota in our Western societies are constantly hungry. Worse yet, when fed with foods suboptimal for their metabolism, they literally take revenge for being put on a diet and start fighting back after a while: the production of toxic metabolites is the microbes' answer to our poor, daily dietary choices and have become an unexpected risk factor for the maintenance of our long-term health.

Impaired human cell functions and starving microbes due to unmet dietary needs are increasingly recognized as the ultimate common ground for the biggest epidemic in human history: diet-related chronic diseases.

Chapter

2

THE "INDUSTRIALIZED" MICROBIOME

A CAUTION LABEL FOR A GLOBAL EPIDEMIC

DOI: 10.1201/9781003212447-2

Humans have an intimate relationship with the microbes covering our body and populating our guts. In fact, we form an inextricably connected composite organism of human and microbial cells. The importance of this special liaison for determining the individual phenotype led scientist Lynn Margulis to propose the term "holobiont"[44]. Our relationship is generally highly mutualistic.
At the most basic level, we provide bacteria with easy access to food and shelter, while the bacteria aid the host in digesting complex food and synthesize essential metabolites, such as vitamins B and K[45].

Over the past two decades, it has become clear that this relationship is far more complex than we have originally thought. Microbes support the development and healthy function of all our organs, influence our metabolism and regulate the immune system which ultimately determines our state of health and the quality of our life. However, with the advent of industrialization of food, sanitation, and modern lifestyle, the relationship with our microbes has begun to change and has become fragile. This perturbation is increasingly associated with the onset and development of chronic human diseases as we will cover soon in detail.

Looking back in human evolution the transformation of the human holobiont began already with the divergence of humans and chimpanzees, marked by a substantial loss of gut bacterial diversity. In fact, humans living in today's industrialized societies harbor the lowest levels of gut bacterial diversity of any primate for which metagenomic data are available[46]. Since then, this bacterial depauperation has accelerated with human adaptation to a modern lifestyle.

DNA sequencing analyses of ancient pieces of evidence such as coprolites (ancient human feces), dental calculus, frozen tissue from permafrost or mummified remains dating back 8000 years provide proof that the diversity of human microbiomes in the gut and mouth have significantly decreased throughout the course of the hominin evolution[47–50].

A study published in 2021 uncovered that the microbiomes of ancient tribes, analyzed from well-preserved paleofeces found in caves in Utah and northern Mexico, were dramatically different compared to fecal samples from people living a modern lifestyle, but closely match microbiomes from individuals of contemporary tribes still consuming a traditional, non-industrialized diet[51,52]. Strikingly, almost 40% of the analyzed microbial species in ancient samples are missing in today's Western guts, indicating an "extinction" of human microbes in modern lifestyle populations.

Even the gut microbiomes of current rural tribe members across different geographical regions including Venezuela, Colombia, Burkina Faso, Tanzania, Papua New Guinea and Central Africa share a common denominator: they are significantly richer and more diverse in their species composition and harbor less opportunistic pathogens compared to the peers in the same region or the United States, for example[53–59]. Instead, Western microbiomes are shown to have around 30% less species, which is accompanied by a large shift in microbial populations from plant fiber-degrading *Firmicutes* to protein-degrading *Bacteroidetes*[53–59].

Studies of captive primates with artificially manipulated diets provide helpful context for understanding human dietary transitions and its impact on the microbiome. Several studies report captive primates consume less diverse, lower-fiber diets compared to their wild counterparts, mirroring the gradual transition to low-fiber diets over the course of human evolution toward today's Western diets. The low-fiber diet resulted in "humanization" of the gut microbiome in the primates, as characterized by the loss of microbial diversity[60,61].

Scientists fear that all the missing microbes found in non-human primates and other rural living individuals but not in the modern guts of our Western societies may become globally extinct over the next generations and become unrecoverable together with their important functions for human health. They suggest that the lost microbes are "symbionts lost in urban-industrialized societies" and fear serious consequences for human health[55].

In order to preserve the microbial diversity important to humanity, leading microbiome researchers, namely Rob Knight, Martin Blaser and Maria Dominguez-Bello along with others, launched the non-profit project the "Microbiota Vault", a global microbial community biobank. According to the scientists, the "Microbial Vault could serve as a model for preserving representative endangered environmental and organism-derived microbial communities"[62–64].

Humanity is losing its microbes, but does it really matter?

MISSING MICROBES – DOES IT MATTER?

Microbiome researchers are concerned about the erosion of the human microbiome in Western societies due to rapid lifestyle modernization, the overuse of antibiotics, exaggerated hygiene measures, medical practices and dietary changes. They hypothesize that "individuals in the industrialized world may be harboring a microbial community that, while compatible with our environment, is now incompatible with our human biology"[65]. Research suggests a direct relationship between a shrinking microbiome and the onset of modern human "plagues", including autoimmune disorders, asthma, type 2 diabetes, heart disease, obesity and cancer[65,66]. The leading microbiome experts, Rob Knight and Martin Blaser, ponder

> are all these distinct diseases independent, or is there
> a common underlying factor? We believe that changes
> in the human microbiota occurring concomitantly
> with industrialization may be the underlying factor.
> The changes involve the loss of our ancestral microbial
> heritage to which we were exposed through millions of
> years of evolution[63].

The strong correlation between microbial loss and the advent of noncommunicable diseases among Western countries has even been announced as a "global microbial biodiversity crisis" by scientists in a workshop held by the esteemed National Academy of Sciences in 2019[67]. The Erica and Justin Sonnenburg lab propose the term "microbiota insufficiency syndrome (MIS)" to describe the loss of microbial taxa and associated functions, which are essential to the maintenance of human health.

> It is our opinion that aspects of our microbial identity
> have gone extinct and that this extinction results in a
> mismatch between our recently adapted microbiota and
> our more slowly adapting human genome … In theory,
> this incongruity in host and symbiont could result in
> society-wide detrimental health effects[68].

Awareness around the "missing microbes" hypothesis was initially created by Dr. Martin Blaser in his same named book He condemns the ubiquitous, inappropriate use of antibiotics leading to a long-term disturbance of our gut microbial ecosystem. Antibiotics do not only act on "bad" bacteria that cause infections but also affect the resident, commensal gut microbiota, our "good" bacteria. He warns against an "antibiotic winter" in which antimicrobial resistance and disease prevail in the future of our societies[69].

Indeed, epidemiological studies show that antibiotic intake early in life comprehensively disrupts the microbial community in the gut and increases the likelihood for children to suffer from asthma later in life[70,71]. An antibiotic-disrupted microbiome community has been observed to take months or even years for recovery, if at all[72]. Preclinical studies noted that antibiotic treated lab animals were prone to ulcerative colitis, type 1 diabetes, adiposity and psoriasis.[73] "Disruption by antibiotics causes a shift (of the microbiome) to an alternative stable state. The full consequence of this is still unclear", as Dr. Relman and his colleague Dr. Dethlefsen of Stanford University raise concerns[74].

A population study from 2016 identified lifestyle and diet-related microbiome depauperation within a population of the same geographical region as a key factor responsible for a higher rate of autoimmune disease type 1 diabetes, as observed among urban Finnish and Estonian children compared with less urbanized Russian kids[75]. Alex Kostic, Assistant Professor at Harvard Medical School and one of the study authors, points out: "We were able to identify specific microbes and microbial products that we believe hampered a proper immune education in early life. This leads later on to higher incidence of not just type 1 diabetes, but other autoimmune and allergic diseases"[52].

Low microbiome richness is also associated with immunological susceptibility for inflammation and chronic diseases. For example, a loss in microbial diversity, was also shown to correlate with a significantly increased infection risk based on data of more than

70,000 patients with type 2 diabetes and obesity in several large studies[76-78]. Also, infants delivered by caesarean section, which are found to have trouble colonizing a diverse microbiota throughout the first years of life, display a weaker immune response than babies delivered vaginally[79]. They are even more likely to suffer from allergies, asthma and diarrhea later in life[80-82].

All these observations indicate an important role of a more diverse microbiome in the development and maintenance of a balanced immune response, an important prerequisite for protecting us against diseases and cancer while at the same time suppressing autoimmune diseases.

The current assumption is that the rich gut microbiome of indigenous people probably reflects the original state of the human microbiome, while modern populations have less diverse microbiomes and are missing an array of microorganisms, which is believed to make them vulnerable to all the chronic diseases ailing our present societies. Encouraged by a vast amount of correlative evidence demonstrating the association between missing microbes and modern diseases, a higher microbiota diversity seems to be generally accepted as health promoting. Endorsing this idea, a biodiversity intervention study showed that being exposed to nature every day increased the commensal microbiota diversity and resulted in more balanced cellular immune response activity in daycare children, important for protection against infections or autoimmune diseases[83]. The study authors concluded that regularly being exposed to nature may be a great prophylactic approach to reduce the risk of immune-mediated diseases in urban societies in the future by increasing the individuals' microbial abundance.

In support of the loss of diversity hypothesis, the prevalence of chronic diseases including obesity, type 2 diabetes, cardiovascular disease and kidney disease was found to be significantly higher in individuals who have taken antibiotics in the last year compared to healthy individuals[84]. Conversely, calling the loss of diversity hypothesis into question, some studies fail to demonstrate a significant difference in the overall microbial diversity and composition associated with a disease state. For example, in a case of colorectal cancer they even detect either no or a reproducibly higher microbiome richness in patients than in controls[85,86].

The lack of strong causal evidence and contradicting results contribute to rising uncertainty and disagreement among the research community concerning the microbiome's involvement in chronic diseases. Apart from the complexity of our microbiome per se, and the short period of time to gather insights about it since the young discipline of microbiome science took off, widely underrecognized factors fueling the growing inconsistencies are the lack of standardized microbiome study methodologies, huge disparities in the common understanding of the microbiome and premature conclusions on causality collectively leading to an accumulation of erroneous results distorting the evidence. These issues represent a massive caveat and limitation for gaining a solid understanding about the microbiome-disease relationship and call for a new perspective in this research field, as we will discuss in the following chapter.

Researchers began to realize they are left with more questions than answers:

> A number of questions remain surrounding conclusions that dysbiosis contributes to the manifestation of disease. First, is the differential abundance of OTUs (operational taxonomical units) between two populations driven by inter-individual variability of the two populations or disease status? Second, what features of microbial taxa contribute to them being identified as dysbiotic?

Finally, are there taxa consistently found to be dysbiotic across chronic disease states, indicative of common, dysbiosis-driven disease processes?[84]

In order to find more satisfying answers and figure out the clinical significance of the microbiome in disease development, we may need to take a step back and rethink what is understood to be a "healthy microbiome" or a "sick (dysbiotic) microbiome" in the first place, as well as what influences the human microbiome to become associated with disease progression or even involved in its onset.

Is a higher microbiota diversity healthier? Are some microbes more important than others? Are there good or bad microbes? Is it really a specific set or abundance of microbial taxa contributing to a disease or other traits? Is the microbiome a causal factor or rather an independent victim and bystander in the spread of chronic diseases worldwide? If so, how does the microbiome trigger diseases and most importantly how can we intervene and prevent this development?

Throughout the book, I aim to shed a little light on these points. I also wish to propose a more simplified perspective on the microbiome in health and disease, which, however is already about to become of tremendous clinical significance in health care systems and of goal-oriented progress in microbiome research.

In fact, it may not be so much a loss of certain species or a loss of its diversity in our guts, but more what results from it: a comprehensive shift in individual microbiota composition leading to a universal loss of its healthy *functions*. Accumulating evidence suggests that independent from the highly individual taxonomy or a low species abundance, the microbiome in patients *behaves* differently than in healthy individuals.

Chapter

3

THE SUBTLE PROBLEMS OF MICROBIOME RESEARCH

DOI: 10.1201/9781003212447-3

Is the Microbiome "Oversold"?

After the accomplishments of the *Human Genome Project* in 2001, the British microbiologist Julian Davies argued that although completing the human genome sequence was a "crowning achievement" in biology, it would be incomplete until the synergistic activities between humans and microbes living in and on them are understood[87]. At the same time, other researchers called for a "second human genome project" that "would entail a comprehensive inventory of microbial genes and genomes" providing the common ground for the national *Human Microbiome Project* (HMP) initiative[88]. The project jumpstarted in 2007 aiming to understand the plethora of microbes living with us and how to "manipulate the human microbiome" to improve human health[89].

Since then, microbiome science took off and soon has become one of the hottest research areas. A growing number of studies seemed to identify a correlation between a loss in microbiota diversity, changes in the gut microbial composition and the development of chronic diseases in humans. In recent years, many studies have indicated that specific features in the gut microbiome are associated with different health markers, and reported correlations between a specific microbial genera abundance and different lifestyle or dietary factors[53,90].

The epidemiological and clinical evidence backs up the array of preclinical analyses of the past 15 years in which scientists caused and cured analogs of chronic diseases in lab animals through microbiome manipulation, attempting to show a role of the microbiome in disease development.

Results from human and preclinical analyses seem to identify a significant correlation between the abundance of certain gut bacteria or a microbiota composition, the bacterial diversity and the etiology of chronic diseases, claiming that chronic disease development may be characterized and even predicted by a specific microbiome "signature".

For example, a comprehensive analysis of a population-based cohort claimed that 17 of 43 diseases showed a significant correlation with at least one microbiota feature[91].

A certain microbiome "signature" has been proposed for patients with pancreatic cancer[92,93], obesity[94], type 2 diabetes[95–97], inflammatory bowel disease[98,99], multiple sclerosis[100] or colorectal cancer[86,101]. It has even been suggested to predict the mortality rate in hematopoietic stem cell transplanted patients[102,103].

Depending on the study, a specific "microbiota signature" is either defined as species composition, a small set of single species, microbiota diversity or its richness. In fact, most of the current evidence claiming a role of the microbiome with human diseases is based on the idea of species abundance or a disease-specific taxonomical composition which turns out to be problematic and misleading. For example, while some studies show an overall microbiota composition but not specific taxa to be related to a disease, which fails to be the case for an individual patient though, others claim the opposite[85,100,104]. This results in a puzzling confusion among researchers complicating the advancement of microbiome research on the one hand but also its translation into the clinic. Currently, large-scale human studies generating a useful understanding of the relationship between microbial composition and diseases remain scarce and are often underpowered to detect robust, replicable associations.

CORRELATION DOES NOT PROVE CAUSALITY

Regardless of the huge disparities in research results, scientists and the health care industries are tempted to draw premature conclusions about a causal role of the microbiome in human disease development based on observed associations. According to psychologist Daniel Kahnemann, this reaction may seem normal for human beings as "our brain is a machine for jumping to conclusions" and can lead us astray. Yet, some scientists find fault with scientists and media having established an "exaggerated causality" thinking merely based on contradicting and lacking, solid human evidence as well as major technical flaws: "Animal-based microbiome research might have played a role in overstating the causal effects of the microbiota in human health and disease" and fear that "Microbiomics risks being drowned in a tsunami of its own hype"[105,106].

In particular, they point out that the reproducibility of microbiome associations across independent studies is low, with additional significant methodological confounders being the DNA extraction step, improper and inconsistent documentation of sample collection and processing, yet-to-be standardized data processing and analysis methods. Also, a lot of studies rely on the analysis of 16S rRNA, an ancient, preserved gene that is reliably found across the bacterial kingdom but does not define specific strains.

> Microbiomes associated with obesity have been distinguished by different ratios of bacterial phyla, which encompass a staggering range of diversity. If this criterion were used to characterize animal communities, an aviary of 100 birds and 25 snails would be considered identical to an aquarium with 8 fish and 2 squid, because

> each has four times as many vertebrates as molluscs. Even
> within a single species, strains often differ greatly in the
> genes they contain

criticizes William Hanage, Associate Professor of Epidemiology at
Harvard School of Public Health, in a *Nature* article[106].

Apart from that, microbiome studies often rely on germ-free mice,
which do not represent the animals' natural state and are typically
unhealthy owing to the lack of microbes in their guts. Therefore,
results are unlikely to predict responses in animals with healthy
microbiomes let alone in humans. In addition, microbiomes in
mice and humans are adapted to different niches with animal
study-based data being hardly translatable to the human
"real-world" situation.

Finally, stool consistency is often referred to as the single personal
factor with huge variation in microbiome composition even among
healthy individuals or within the same stool sample. In a systematic
review, Dr. Jens Walter and colleagues found that 95% of published
studies (36/38) on Human Microbiota-Associated (HMA) rodents
reported a transfer of pathological phenotypes to recipient
animals, and many extrapolated the findings to make causal
inferences to human diseases.

> We posit that this exceedingly high rate of inter-species
> transferable pathologies is implausible and overstates
> the role of the gut microbiome in human disease. We
> advocate for a more rigorous and critical approach
> for inferring causality to avoid false concepts and
> prevent unrealistic expectations that may undermine
> the credibility of microbiome science and delay its
> translation[107].

Regardless of the critics, the number of studies mentioning "microbiome" or "microbiota" in their title or abstract grew from 11 in 1980 to over 13,000 in 2018, making it hard to slow down the excitement and gain back seriosity[108]. More than 20 years after the first paper detailing the immense diversity of human microflora in 1999, microbiome research has gained huge popularity and is supported by millions of dollars in governmental and private funding signaling the public official significance[109,110]. Interesting findings and promising but premature results have been soaked up hungrily by the public and lay audience with high hopes. Industries have smelled the enormous business opportunity reflected by predicted microbiome market growth rates of over 20% each year to reach 1.6 billion USD until 2028, while microbiome-related patents have hardly increased implicating a lack of meaningful progress amid the noise[109,111].

One editorial in the journal *Nature* demands that "it is time to further strengthen the scientific basis from which microbiome translation can grow".[108] Harvard professor William Hanage writes against the hype and appeals to microbiome researchers for asking crucial questions before drawing any conclusions: "Can my experiments detect differences that matter? What is the mechanism? Is the experiment reflecting human reality? Could anything else explain the results?"[106]

Jonathan Eisen, a microbiologist and blogger at the University of California, even bestows awards for "overselling the microbiome"[112]. Apparently, he finds no shortage of worthy candidates.

Is Microbiome Science Heading in the Wrong Direction?

Despite thousands of microbiome-related studies, the low concordance between human studies limits the capacity to identify causal relationships between our microbiome and disease pathology. It appears that microbiome science is not making the desirable progress that could be anticipated regarding the immense governmental priority dedicated to it, backed by millions of dollars of funding. What is going on?

It all starts with terminology. Apparently, there is no existing clear-cut, universal definition for "microbiome" agreed upon in the scientific community. A few review articles have previously defined the term, but mainly as sidebars, however, no clear definition has been published or enjoys universal consent.

Prof. Julian Marchesi of Imperial College London explains: "The misuse of terms such as microbiome, microbiota, metabolomic, and metagenome and metagenomics among others has contributed to misunderstanding of many study results by the scientific community and the general public alike".[113] How can the scientific community push microbiome science in a joint effort into the right direction if it is not even sure what the research subject exactly is about? The lacking definition of the term "microbiome" not to mention the even less developed understanding of the term "healthy" microbiome, apparently creates some bewilderment among the research community which contributes to the cumbersome progress in defining a microbial disease "signature".

In addition to ill-defined terms, some inconsistencies may also be attributed to confounding factors that significantly influence data yet are widely neglected. For example, researchers identified several lifestyle factors and physiological traits in humans impacting the microbiome, particularly alcohol intake and bowel movement frequency that may strongly confound microbiota analyses but are hardly taken into account by most studies[114].

Surprisingly, a lot of research still underestimates or even neglects a critical factor for gaining a deeper understanding about the relationship between the human microbiome and chronic diseases: the microbes' behavior.

> A big gap remains between our current understanding of microbial patterns and our ability to apply such knowledge to develop treatments. Most studies have focused on identifying the species composition of the human microbiota and, to a lesser extent, the 'functional potential'. However, the actual activity of the microbiota, such as the specific molecules that are being produced or degraded at a given time, is rarely determined in present studies

as microbiome researcher Dr. Veronica Llorenz-Rico points out in an article on the *integrated Human Microbiome Project* (iHMP), the second phase of the influential *HMP*[115].

The iHMP studies the human-microbe interaction in diseases and gathers an enormous amount of joint human-microbe data which may be useful for understanding and treating diseases. However, scientists raise concerns that "even bar-raising studies such as those conducted by the iHMP have weaknesses. In the past few years, key factors that affect microbiota composition have been identified, and although some, including age, were considered in the analyses, several other potentially crucial factors, such as diet or taking medication, were not"[104].

Even further, they find fault with some important details of the iHMP's technical approach to study the microbiome, which seems to be representative for the current approach of microbiome research in general: bacteria species rather than its function are still the main focus among scientists, throughout the medical community and diagnostics[115]. Also, most of the project's data are cross-sectional and mostly depending on one sample per individual at a given time point with small sample sizes leading to a lack of data, unlikely to be sufficient for drawing a valid conclusion about the microbiome's role on a disease development. Furthermore, there is insufficient joint profiling of human and microbial genes, their gene expression patterns and proteins of interest for specific diseases and most of the time generated data are provided in relative units, so bacterial abundances are expressed as a proportion of the total microbial population, rather than as the number of cells, which however may be a relevant information considering the raised awareness around our "shrinking human microbiome".

After all, one of the first steps for gaining a deeper understanding about the microbiome – disease relation is characterizing the gut microbiome more thoroughly by redefining:

> what is a "healthy microbiome" and what is a "dysbiotic" or "disease-associated microbiome"?

Chapter

4

THE GUT MICROBIOME
A NEW PERSPECTIVE

DOI: 10.1201/9781003212447-4

Our Microbiome Is a Metabolic Organ

In 1974, Lewis Thomas noted in his book *The Lives of a Cell*: "We are not made up, as we had always supposed, of successively enriched packets of our own parts. We are shared, rented, occupied …. without them we would not move a muscle, drum a finger, think a thought"[116].

The surprising wealth of microbial diversity and its influence on human bodies led scientists to propose the concept of a "meta-organism" or a "holobiont": all eukaryotes, including humans, must be considered together with their microbiota as an inseparable functional unit[110,117]. Since the 1970s, the holobiont concept of our human body as a cell composite consisting of 10 times more microbes than human cells has been widely accepted in research and the public, however, in 2016 these numbers were revised after which the human – microbe holobiont turned out to share its existence cell number-wise in fairly equal parts: 38 billion non-human cells and 30 billion human cells[7]. So, 43% of our human body is composed of human cells while still more than half of it consists of microbial cells.

Who Is Our Microbial Self?

The first definition of the term microbiome was proposed in the late 1980s by J.M. Whipps while working on the ecology of rhizosphere microorganisms[118]. Whipps and his team of researchers described the "microbiome" as a combination of the words "micro" and "biome", naming a "characteristic microbial community" in a "reasonably well-defined habitat which has distinct physio-chemical properties" as their "theatre of activity"[118]. This more than 30-year-old proposed definition has since often been reduced to a minimal definition of the "microbiota as a community of commensal, symbiotic, and pathogenic microorganisms" among the research community. The Merriam Webster publishing platform even proposes two microbiome definitions: one describing the metagenome and the other is the community of microorganisms, yet still fails to capture the host and the environment as an integral ecological component of the microbiome, rather than an independent entity[119]. Currently, Whipps' definition from 1988 seems to be the most appropriate as it already includes the distinct properties and functions of the microbial community and considers its interactions with its environment, resulting in the formation of specific ecological niches.

To date, bacteria are the most studied members of the microbiome, hence one can easily get the impression that the term "microbiome" refers to bacteria only. However, apart from bacteria the community of microorganisms consists of a tremendous variety of other microbes, which are still neglected and understudied members of the microbiome. Most microbiome researchers agree that bacteria, archaea, fungi, algae and small protists should be considered, yet whether phages, viruses, plasmids and mobile genetic elements are part of the family, too, remains controversial[113].

While microbiome science is still in its infancy of determining which role they play in the "theatre of activity", a clear differentiation between the terms "microbiome" and "microbiota" may provide a good starting point to make progress in finding out which role the microbiome plays in human health.

Generally, the microbiota is defined as the collection of living microorganisms present in a defined environment.[113] Phages, viruses, plasmids and free DNA are usually not considered as living microorganisms, hence they do not belong to the microbiota. The original proposition for the term "microbiome" by Dr. Whipps, however, includes not only the community of the microorganisms, but also their "theatre of activity". This "theatre" involves the whole spectrum of molecules produced by the microorganisms, their metabolites, which importantly reveal their function. In an attempt to resolve the discussion about a universal terminology some researchers propose that "all mobile genetic elements, such as phages, viruses, and extracellular DNA, should be included in the term microbiome, but are not a part of microbiota"[120].

Regardless of the ongoing discussion about a universal terminology, the most crucial aspect for our relationship with our microbes is the way in which we communicate with each other. We need to "talk" and interact in order to become and maintain a healthy human "we/us". Researchers uncovered that nearly half of our blood metabolites are of microbial origin[9]. The metabolites are the microbiotas' language, their way of having a conversation with us. "Despite the widespread host genetic effects on blood metabolites, the gut microbiome might play a role on the systemic metabolism that is independent from the host genome", as London-based researcher Dr. Alessia Visconti argues[9].

Our microbiota provides us with genetic and metabolic attributes, depending on the substances they produce, which we have not been required to manifest on our own, including the ability to harvest otherwise inaccessible nutrients. For example, the human gut microbiota may contain up to 60,000 carbohydrate-degrading enzymes, while on the contrary the human genome, harbors only 17 enzymes that are involved in carbohydrate digestion.[121] Consequently, our human digestion system relies heavily on the help of our microbes when it comes to digesting food.

Besides, a wide range of other functions are being discovered and elucidated, from its influence on host gene expression, angiogenesis, postnatal intestinal maturation to immune system development and training. The microbiota even performs "detox" activities, as it metabolizes substances that may be harmful to the human host such as heavy metals, pesticides, persistent organic pollutants, artificial sweeteners, nanomaterials and other food additives ingested and modulates their toxicity[122]. For example, the microbiota may modify medical drugs considerably affecting the efficacy and toxicity, a side fact that is still left unconsidered in clinical research studies and which we will cover in later sections.[123]

Due to the substantial metabolic performance, the distal human intestine has been named an "anaerobic bioreactor" programmed with an enormous population of bacteria and other microbes producing substances that impact human health.[124] In a scientific workshop carried out by the National Academy of Sciences in 2006, scientists concluded, that our gut microbiota can be pictured as an "exquisitely tuned microbial organ placed within a host organ"[125]. According to the scientists, this microbial organ is to be composed of different cell lineages with a capacity to communicate with one another and the host and consumes, stores and redistributes energy. It acts as a "master of physiological chemists" mediating physiologically important transformations and employing a broad range of other strategies to manipulate host genomes[125].

In 1992, the Italian Prof. Velio Bocchi suggested that "collectively, the flora has a metabolic activity equal to a virtual organ within an organ"[126]. In fact, its metabolic rate rivals that of our liver consuming 250–300 kcal each day.[127,128] The microbial metabolites influence and maybe even control our cellular activities in our organs. Dr. Michael Fischbach, a bioengineer of Stanford University, suspects

> They (the metabolites) are naturally present in the body at concentrations similar to those of therapeutic drugs.

So there is no question about whether these molecules
end up permeating the host – they do. There are dozens
of them, and we feel it is warranted to take the time to
treat these molecules almost like a classic pharmacology
problem[129].

The most crucial "pharmacological" substances produced by
our gut microbiota for preventing chronic disease development
have been found to be the short chain fatty acids (SCFAs) such
as propionate, acetate and butyrate, generated from dietary
fiber present only in plant foods. Importantly, butyrate feeds the
intestinal epithelial layer cells important to maintaining a healthy
intestinal barrier function[122]. As microbial-produced SCFAs are the
key energy source for our intestinal epithelial cells, around 90% of
the molecules are eaten up by colonocytes and enterocytes, while
merely 10% enter the blood system[130–133].

An intact intestinal epithelial layer covered by a thick mucus
layer is essential for human health while its breakdown and
"leakiness" is proposed as an initial event in disease development,
as we will discuss soon. Mucus is a viscoelastic gel that protects
the intestinal epithelial cell layer, separating it from the lumen
content. It can literally be considered a "firewall", a dynamic
barrier, that carefully selects what crosses the line. It is permeable
to gases, water and nutrients, but not to microorganisms or some
food components. For a long time, the intestinal barrier was
simply considered as a physical barrier between the environment
(including the lumen content) and our blood system, however,
it also provides the vital habitat for the human gut microbiota it
needs to thrive and flourish.

Although the majority of the SCFAs are consumed by our gut lining
itself, the remaining homeopathic, micromolar concentrations

entering the periphery may be enough to influence the processes in our body: glucose homeostasis, lipid metabolism, appetite regulation, immune response and even brain activity. The low concentration as well the fact that short chain fatty acid are not very stable substrates or designed to persist for long, imposes a challenge on diagnostics to detect them. Some researchers argue that it is unlikely to reach physiologically relevant levels in the periphery and that many of the beneficial effects of short chain fatty acids, for instance, are likely to be exclusive to the colon, where the metabolites are estimated to be 1000-fold higher than in the periphery.[131] Opposing this hypothesis, an abundance of cellular receptors for these microbial metabolites, however, have been detected in nearly all human tissues: fat tissue, bones, liver, pancreas, kidney, small intestine, immune cells, the peripheral nervous system and even the brain, indicating a considerable role of these metabolites in maintaining cellular processes in our organs[134,135].

Only few human studies exist proving a direct influence of short chain fatty acids on human body functions or even disease protection. The majority of patients suffering from a chronic disease, or a gastro intestinal disorder, are shown to have less SCFA-producing bacteria and especially butyrate in their gut lumen, resulting in a breakdown of the intestinal barrier and a leaky mucus firewall which is suspected to correlate with and aggravate inflammation in these patients, e.g., inflammatory bowel disease, type 2 diabetes, celiac disease, inflammatory bowel syndrome or colorectal cancer[85,96,136–139]. In contrast, a treatment with or a stimulation of a healthy SCFA concentration is associated in some studies with improved intestinal barrier function, the suppression of colorectal cancer cell proliferation *in vitro* and clinical improvement in systemic inflammation in healthy and overweight people[140–142].

Ample preclinical evidence on lab animals continues to paint a vivid picture of how SCFAs possibly impact all essential physiological processes in humans. Apart from metabolic and immunological processes, our behavior and brain function may fall under our microbes' influence too. Microbial metabolites are shown to cross the blood-brain barrier to influence cognitive functions such as learning and memory building, improve sleep and fat and mitochondrial energy metabolism via a gut-brain-neural circuit[130,143,144]. Besides SCFAs, selected bacteria also synthesize neurotransmitters, such as gamma–amino butyric acid (GABA), serotonin, catecholamines and histamine, which all together affect the central nervous system (CNS) via the enteric nervous system (ENS) and enterochromaffin cells ultimately affecting mood, eating behavior or perception of pain as has been observed in animal studies[143,145–147]. In humans, a higher concentration of a particular microbial metabolite, called equol, which bacteria make from dietary soy, is associated with significantly less white matter lesions (WML) in the brain, indicating a lower risk for dementia[148].

The constantly accumulating pieces of evidence validate the significant effect of certain microbial metabolites rather than a taxonomical composition on human health maintenance and disease development. These metabolites act like hormones being as influential on the human body processes as thyroid gland hormones. The gut microbiota, however, is only able to produce these hormones if it has sufficient substrate available for metabolic turnover: components from the food we eat, but also molecules shared by other microorganisms in the local gut environment community. Hence, the type of food and the local microbiome environment in the gut, the niche, are pivotal prerequisites for the synthesis of metabolites that, depending on the substrates provided, can be either health maintaining or disease promoting.

Importantly, to produce beneficial molecules especially short chain fatty acids, bacteria almost exclusively rely on their favorite substrate: dietary fiber.

Dietary fiber is the most crucial ingredient in plant-based foods, that our gut bacteria thrive on and consequently serve other human organs with "therapeutic" metabolites. Consequently, depriving the microorganisms of their most vital source of energy while simultaneously feeding them substrates suboptimal for their metabolism, they begin to starve. The architecture of the microbial ecosystem shifts from a health-promoting, eubiotic state towards a dysbiotic state releasing large amounts of non-beneficial metabolites. As a result, the human gut microbiome can behave in a "toxic" manner with detrimental effects on the health of the human host.

Current evidence suggests that a toxic microbiome rather than a specific taxonomical composition may be the molecular breeding ground for the development of chronic diseases and cancer.

FUNCTIONAL *OMES*: METABOLITES OVER SPECIES

Our gut microbiome is far from a random assembly of certain microbial species. It is rather a complex, well-defined ecosystem following its own rules.

In his book, *The Lives of a Cell,* biologist Lewis Thomas describes vividly: "The bacteria are beginning to have the aspect of social animals ... They live by collaboration, accommodation, exchange, and barter. They, and the fungi, probably with help from a communication system laid on by the viruses, comprise the parenchyma of the soil ... They live on each other"[116].

A disruption of this lively ecosystem in our guts has been observed to be strongly related to the development of chronic diseases.

Studies on coral reefs trying to elaborate natural ecosystem workings and the consequences of a disruption analyzed that a species biodiversity loss exponentially decrease the ecosystem's stability and functioning, a finding which may be applicable to other ecosystems such as the human gut microbiome as well[149].

Stability and function of an ecosystem is the result of a lively and smooth interaction between the systems' members. No species thrives on its own. "No man is an island entire of itself" as the English poet John Donne described the vital connection between all of humankind already in 1624[150].

Concerning the human gut microbiome, the mutual interaction between microbes and hence their functionality is highly dependent on local microbial communities and their spatial arrangement in functional units, so called "niches". These countless functional entities are arranged in a defined three-dimensional architecture that is necessary to effectively interact with each other. For example, they feed on metabolites secreted by microbial neighbors, as isotope studies in mice and humans have found, but they also exchange genetic material[151,152].

These mutual interactions may convey the illusion of a harmonious relationship, which however is not so harmonious after all, since the ecosystem participants constantly compete for space and nutrients[125]. The local selection of microbes and metabolites determines the population structure in that niche, and by competing and cooperating, populations from adjacent niches influence the overall composition. The interplay between collaboration and deadly competition is critical for strengthening the stability of the microbial ecosystem: a "costly" yet productive cooperation[153].

When looking at the microbial architectural "building" on a broader scale the bacterial density increases along the small intestine and rises to 10^{12} bacteria per gram of colonic content in the large intestine, ultimately contributing to 60% of fecal output[154]. Not only the longitudinal microbiome structure differs along the gut from stomach to rectum, but also its vertical composition in a cross-section of the intestinal gut wall. Epithelial surface-adherent, intestinal crypt located, and luminal microbial populations vary strongly as well as the ratio of anaerobes to aerobes which is lower at the mucosal surfaces than in the lumen[154].

Although the gut microbiome is a highly organized organ, its cellular communities are arranged less rigid and more loosely compared to the microbes on static surfaces such as the teeth[155,156]. This microbial architecture provides the stage for the concert of interacting bacteria, fungi, bacteriophages and viruses and is a framework for their cross-feeding activities and production of metabolites.

Precise microbiome mapping is still a new field of research, however a growing body of evidence suggests that the spatial structure has a powerful influence on the stability of mutualistic relationships between members of the microbial community on the one hand and the wellbeing and health of the holobiont on the other hand[155,157,158].

A study published in *Nature* in 2019 found, based on preclinical data, that only a microbial "consortium", a whole community but no single species "work in isolation", has the power to enhance anti-tumor immunity: "It's an assembly of microorganisms that can collectively impact genome stability and immune function. Assembled in certain consortia, they had a much larger effect"[158].

Every human being has a quite unique microbial species composition that varies markedly between individuals and even intra-individually from day to day depending on the choice of foods and other environmental exposures. Although there is extensive interpersonal variability in microbial composition, importantly all the different microbiota in different people have core functionalities. Unrelated people and even twins are shown to share merely 43% of the microbial species, however the metabolites in their blood and feces are more than 80% equal[9,159,160].

Only the defined spatial arrangement of microbes allows them to perform similar roles in the "theatre of activity" that either benefit us or wreak havoc on our health.

Unlike most of the current research still does, some researchers are starting to become aware of the importance of the microbial metabolic potential rather than exactly which microbial species is involved: "These surprising observations underline the importance of studying the microbial metabolic potential rather than focusing purely on taxonomy. Future treatments … should optimally target functionally related microbial communities rather than single organisms"[9].

Noncommunicable diseases and cancers are rarely related to single pathogenic microbes. Only 16% of all cancers and even less of all chronic disease cases can be attributed to specific pathogenic infections, while the majority is correlated to an overall disruption of the microbial ecosystem[161,162].

Historically, the study of diseases has been approached from a "one microbe-one disease" viewpoint. Viruses, eukaryotes and bacteria were studied under conditions in which they were believed to cause disease. Throughout the twentieth century, single-pathogen diseases such as measles and tuberculosis (TB) fell to advances in medicine and public health, while the rates of multiple sclerosis, Crohn's disease, type 1 diabetes and asthma have skyrocketed. "This raises questions about the unintended consequences of attempts to eradicate single pathogens such as TB and measles on the rest of the microbiome" as Rob Knight raises concerns[163].

Regardless, a global joint effort was started years ago to eradicate the gastric bacterium *H. pylori* from the human microbiome, shown to be involved in the development of stomach ulcers in 1 out of 100 people colonized with the bacterium[164]. Intriguingly, more than 80% of the world's population is asymptomatically "infected" with *H. pylori*. Even its explorers, the 2005 Noble prize laureates Barry Marshall and Robin Warren, postulated that the organism may in fact play an important role in normal gastric ecology since *H. pylori* has been reliably present in the human gastric niche for more than 58,000 years and so far, no other single species turns out to be so dominant in the human body[165,166]. Its ability to promote growth factor expression is assumed to be even essential to the development of the host early in life[167]. Supporting these observations, the eradication of *H. pylori* is associated with a disruption of the hormone system, increasing the odds for weight gain, esophageal cancer and even asthma in humans[168–172].

Hence, although in very low numbers, even opportunistic pathogens such as *H. pylori* are standard members of the microbiome and ultimately play a role in stabilizing the spatial architecture of the microbial organ. Dr. Jose Clemente points out: "Just as the 'one gene-one enzyme' outlook proved to be an oversimplification that failed to explain complex phenotypes, we are now starting to appreciate the fact that some diseases might result from dysbiosis rather than the presence of a single disease-causing microbe"[173].

The microbiome can be destabilized by different factors which results in a dramatic functional change or even loss of its healthy functions, indicated by the type of microbial metabolites in the human blood and feces. An unhealthy change in the microbiome's behavior is often described as *dysbiosis*. Healthy individuals seem to share a similar metabolome while patients suffering from a disease could be identified by a distinct, disease-specific "toxic" metabolic signature. Importantly, the shift from a health-supporting to a disease-associated microbiome can occur independently from the gut microbiota composition, which is a significant observation as this may simplify diagnostics tremendously, as we will soon discuss, since each person's microbiota is very individual. For example, although certain plasma metabolites in patients with metabolic syndrome explain the degree of insulin sensitivity, these metabolites are not linked to the individual gut microbiota composition[174]. Similarly, unlike in healthy individuals, microbial specific plasma cells and bacterial DNA were found in the central nervous system of multiple sclerosis patients and in the joints of rheumatoid arthritis patients which correlated with disease relapse[100,175–178].

"I don't think the bacteria move, but their metabolites do", as neuroscientist Patrizia Casaccia confirms the compelling concept of metabolites as a predominant linking factor between diseases and anomalies of the gut microbiome[175]. In fact, this metabolomics approach is opening an intriguing possibility for developing disease-prediction tools in the near future. Researchers were already able to reproducibly identify cancer biomarkers based on merely microbial metabolic pathways in the colorectal cancer patients' feces[86,176,85].

While the metabolome differentiates healthy people from patients, the highly individual microbiota composition may be an identifier for a specific person or its lifestyle: a microbial "gut print".

After one year of taking a microbial sample from different body parts in various persons, in almost 90% of the samples researchers still were able to identify the matching donor in an anonymous study and were able to uncover details about that person's health, diet and ethnicity[179,180]. In another study, microbial traces from hygiene products or a person's mobile phone provided information regarding a subject's personal habits and individual lifestyle[181].

These results raised serious privacy issues among researchers. Although it may still be challenging to do anything with the microbiome data from a single study, the growing number of studies and open-source databases such as the HMP likely increase the risk of privacy violation. Curtis Huttenhower, a computational biologist at the Harvard T.H. Chan School of Public Health in Boston, Massachusetts, who led one of the studies stressed that "as the field develops, we need to make sure there's a realization that our microbiomes are highly unique"[180].

An individual microbiome sample may accidentally reveal more about oneself than one wishes to share.

The complex 3-dimensional architecture of the gut microbiota is hard to capture by current diagnostic tools and the huge diversity in inter-individual microbiomes makes the replication of bacterial species-level associations with host phenotypes scarce. It is therefore questionable if current species-level diagnostics on its own are longer suitable for providing useful information to researchers or in the clinic. The currently popular stool analytics used in clinical diagnostics are majorly based on bacterial species compositions, hence the tests fail to capture the relevant intestinal microbiome architecture and the vast array of metabolites traveling throughout the body. Therefore, the results are unlikely to provide meaningful information about the real function of the microbiota in that patient and are deemed unsuitable for a disease diagnosis, its management or even prediction. For instance, the German Gastroenterology Association DGVS ("Deutsche Gesellschaft für Gastroenterologie, Verdauungs- und Stoffwechselerkrankungen") even strictly advises against common stool assays to examine one's gut microbiome as "these currently lack the scientific basis" and are "expensive and pointless"[182].

The English zoologist Thomas Bell cuts to chase of the importance of the microbiome's community architecture for its function in the human body when quoting Douglas Adams, the author of the famous book *Hitchhiker's Guide to the Galaxy*:

> If you try and take a cat apart to see how it works, the first thing you have on your hands is a nonworking cat', and this is an important issue when endeavoring to conserve a microbiome sample. How would you distinguish a 'whisker' from a 'heart' and assess what a component does in the microbiome and its relationship with the total functionality of the whole system? For example, the very nature of a crop-related microbiome changes when one changes the crop variety, the management practices, or adds or eliminates microbes[183,184].

In addition to the clinical hurdles the current approach to study the microbiome brings along, some scientists argue, it is difficult to reveal important functional elements of the gut microbiome solely by metagenomics drawn from fecal samples. They propose a more advanced "functional 'omics', combining metaproteomics and metabolomics, that should also be involved in the study of gut microbiome" in line with the gaining ground concept of "function first, taxa second"[185].

Apart from the conceptual flaws blurring the relevant findings of microbiome research, blatant technical issues pile up, ranging from inadequate sample collection and nonconformational preparation to inconsistent definitions and analysis of relevant parameters[186,187].

For example, in current studies there is hardly an association between the detection of butyrate-producing bacteria species in the gut, the butyrate concentration measured in the feces and how much and if metabolites enter the bloodstream or actually result in a clinically meaningful effect[188]. Dr. Serena Sanner of the University of Groningen and her colleagues complain that "fecal butyrate is a poor proxy for butyrate production and absorption. Besides that, plasma butyrate is challenging to measure due to unreliable assays"[188].

Considering the disparities in microbiome research, a collective paradigm shift may be due, sparking a transition from the still prevailing concept of specific microbes that influence health towards the overall *functionality* of the microbiome that may however be quite independent from single microbes.

FINDING A NEW DEFINITION FOR A "HEALTHY" MICROBIOME AND "DYSBIOSIS"

As the scientific community has no consent yet on a universal definition of a *healthy* microbiome: How are we supposed to understand the even higher complexity of a dysbiotic microbiome in different diseases?

While research agrees on the existence of a "core microbiome" at the very imprecise phylum level (consisting primarily of Bacteroidetes and Firmicutes), there are significant disparities between healthy people at lower taxonomic levels, which poses a major

obstacle on managing the challenge of finding a description for a
healthy microbiome.

In 2018, an international panel of 18 experts came together at the
International Cancer Microbiome Consortium to find a consensus
about pressing issues present in microbiome research including a
more precise definition of the microbiome.

In line with accumulating evidence, the expert panel proposes that
a "functional approach is of more utility in discussing normality
and dysbiosis"[189]. According to the panel, a health-associated gut
microbiome has several core features:

> It is diverse, resilient to short-term environmental
> pressures and has sufficient plasticity to adapt to the
> benefit of the host in the face of longer-term pressures.
> The health-associated microbiome should synergise
> with the host to drive beneficial immune responses and
> metabolic mutualism. Finally, the microbiome should
> have a tumour-suppressant effect on the host. Departure
> from these core features can be considered dysbiotic
> and may have the potential to incite or sustain cancer. It
> should be noted that the health-associated microbiomes
> of other niches will have different core features.[189]

Conversely, they acknowledge that dysbiosis is likely host specific
and disease specific – a microbiome may be dysbiotic in one
individual, but not in another and may promote one pathology, but
not another.

> Therefore, dysbiosis is not an absolute and cannot
> be defined outside of the context of the host and the
> disease in question. We propose that dysbiosis should be

considered a persistent departure of the host microbiome from the health-associated homeostatic state (consisting of mutualists and commensals), towards a cancer promoting and/or sustaining phenotype. This dysbiosis is specific to the individual and thus can only be defined by prospective longitudinal analysis[189].

Essentially, a collective paradigm shift requires a progressive, fresh definition of a disease-associated, dysbiotic microbiome which emphasizes the *functional* shift towards a toxic metabolic state as a result of a disrupted individual microbial ecosystem in the gut.

Leading microbiome researcher Rob Knight uses an apt metaphor in his book *Follow Your Gut* to describe the process of a microbiome becoming dysbiotic and toxic:

> What's particularly interesting is that the microbes in patients do not appear to be behaving normally: their metabolism is off; they're eating and secreting different chemicals. We don't yet know if this altered behavior is caused by the body's immune response or if microbes are at fault. Your immune system does not so much keep lists of good and bad microbes as it concerns itself with good and bad microbe behavior. Your immune system is not the FBI conducting a manhunt for John Dillinger. Instead, it's the guard in the bank who freaks out and opens fire when somebody leaps the counter and starts stuffing money into a sack[190].

The central question is: Who or what instigates the microbes in our gut to become criminal and rob the bank in the first place?

Chapter

$$\boxed{5}$$

SHAPING THE MICROBIAL BEHAVIOR

DOI: 10.1201/9781003212447-5

DIET: THE MASTER EDUCATOR OF THE GUT MICROBIOME

The microbiome is a plastic organ composed of a vivid, interacting community consisting of roughly 38 trillion bacteria besides countless other non-bacterial microbes.[7] "Plastic" means the microbiome constantly reacts to external stimuli and adjusts to the permanent environmental pressure we are facing by changing its composition and hence its function.

Where we live as well as our individual lifestyle have the most profound influence on the microbial behavior. Particularly, lifestyle factors such as alcohol and drug intake, antibiotic treatment, sleep, physical activity, diet and environmental exposure to chemicals have been shown to be more powerful in shaping the microbiome than human genes.

For example, analyzing the microbiomes of genetically unrelated people of distinct ancestral origins that share a common environment tell us that the gut microbiome is not associated with genetic ancestry and even host genetics play a minor role in determining the microbiome composition[191]. Based on a genome-wide association study (GWAS), a Chinese cohort study calculated the heritability of the microbiome's α-diversity to be only in the range of 3.5–10.3%[192].

Environment Beats Genetics

Although a person's geographic residence and their direct environment leave a significant imprint on the microbiome, as the population data from three continents revealed (MetaHIT (European), HMP (American) and Chinese (diabetes cohorts), the individual lifestyle, however, is an even more dominant force over genes and geography[193].

An analysis of fecal microbiota composition and functional profiles in two coexisting populations in the Central African Republic with differing lifestyles, the BaAka and Bantu demonstrated significant differences in the abundance of specific microbes and microbiome functionality between the populations. The Bantu community incorporates some Westernized lifestyle practices compared to the BaAka who follow a hunter-gatherer lifestyle. The BaAka may therefore represent ancient humans, with their Bantu neighbors representing a transition to the Western lifestyle. The Bantu microbiome lacked traditional bacterial groups, showed significant differences in the relative abundance of taxa at the phyla level compared to the BaAka overall diminished abundance, and their microbes were metabolically trained to digest chemicals, food additives, simple sugars and drugs, similar to the Westernized microbiomes of Americans[58].

Indeed of all the environmental factors, diet turns out to be one of the most profound, yet still underrated, impact factors for shaping the gut microbiome, surpassing in importance other variables such as ethnicity, sanitation, hygiene, geography and climate.

Our gut microbiota serves as a filter for everything we ingest. Specifically, food can be considered a foreign object that we take into our bodies in massive quantities every day and the microbial communities influences how we experience those meals. By our food choices we can significantly shape our microbial behavior, determine our digestive process and most importantly influence the state of our health in the long term.

Although one might conclude that different daily food choices change the microbiome fundamentally every day, one must not forget that the microbiome is a robust, natural ecosystem that is inherently stable for a long time unless it is disrupted by a strong disturbing factor or is under constant pressure of some strong external forces, that nudge the system towards a new equilibrium.

Up to 90% of the adult human microbiota has been shown to be remarkably persistent in its overall composition and the abundance of taxa on the phyla level for months or even up to a year, meaning that persistence pays off when deciding to opt for a healthy diet change[194].

A cross-sectional study in healthy adults indicated that predominantly long-term diet patterns influence the fecal bacterial communities, with the most striking effects seen in meat eaters as opposed to vegetarians[195]. A low-fiber, animal food-based diet in the study participants was associated with the bacterial composition of the enterotype *Bacteroides*; while conversely, a diet including more carbohydrate, fiber-rich plants resulted in *Prevotella* enterotype-associated bacterial communities[195]. Although the bacterial classification into enterotypes based on genera in microbiome research is of controversial clinical significance today, these early analyses at least created a strong awareness of how profoundly our daily dietary choices can shape the makeup of our gut microbiome above all other environmental factors and genetics.

The long-term imprint of fundamentally different dietary habits on the microbiome was also demonstrated when comparing the fecal microbiota of healthy western European and healthy rural African children. The African group, which ate twice the amount of dietary fiber compared to the European children that were raised by following a typical Western diet including animal protein and processed foods, showed a more diverse microbiome, a four-fold increased production of health-associated short chain fatty acids (SCFAs) and a higher dominant phyla proportion of Bacteroidetes to Firmicutes (the opposite was shown to correlate with obesity in other studies)[53,196,197]. In addition, similar results have been obtained from studies in rural living people in Colombia compared to their urban fellows[56].

Although long-term dietary patterns, particularly animal protein based or plant based, were shown to have significant impact on the overall microbial composition, interventional studies demonstrate a detectable variation in the microbiome metabolism, and on the lower species level from the baseline after exposing individuals to a diet very different from their habitual, long-term diet for only days or a few weeks. For example, controlled feeding studies show reveal that the microbiome metabolism changed detectably as early as 24 hours after switching individuals to a high-fat/low-fiber or low-fat/high-fiber diet, respectively, while the overall composition on the broader enterotype level remained stable during the short period of the study[195,198]. Similarly, switching rural Africans to a "Western" high-fat, low-fiber diet for a few weeks resulted in an increased bile-acid synthesis and a higher abundance of the bacterium *Bilophila wadsworthia*, a sulfite-reducing bacterium whose production of hydrogen sulfide can lead to acute inflammation in the intestinal epithelial cells and is associated with an increased colorectal cancer risk[34]. In contrast, switching African Americans from a standard low-fiber American diet to a high-fiber diet stimulated their microbiota to produce health-promoting short chain fatty acids[34].

We can literally nudge the microbiome metabolism to switch from acting health promoting to disease fostering and vice versa in as soon as a couple of days.

The bacterial metabolism reacts faster and more finetuned than the microbial species architecture to environmental changes which happens mostly independent from species composition. A profound restructuring of the very individual microbiome community in contrast seems to require a more persistent influence, most importantly our food.

In order to survive and maintain the ecosystem, the gut bacteria are primary dependent on an abundance of dietary fiber.

But what happens if the microbiota is deprived of its most vital energy source?

STARVING THE MICROBIAL SELF: A FOUNDATION FOR DEVELOPING A CHRONIC DISEASE

Every living organism exhibits metabolism. Every living organism requires energy. Every living organism must eat.

Although some members of the microbiome are still being debated whether they count as living microorganisms, the trillions of bacteria in our body are shown to be one of the highest metabolically active organs, rivaling the human liver[126].

Since most cells of our microbial self have the highest turnover rate of all organs in the human body, many times faster than dividing human stem and progenitor cells, they require a lot of energy to perform at their best[199,200]. A lot of dietary fiber.

Fiber is composed of mostly carbohydrates contained in all plants as a major structural element while none or merely neglectable amounts of fiber is present in animal tissue[201]. Although fiber is considered a nutrient composed of a complex mixture of dietary carbohydrates, our small intestine lacks the necessary enzymes to digest it. In fact, our human genome harbors only 17 glycoside hydrolases[121,202]. Instead, our digestive system relies heavily on our gut microbiota to take over the fiber breakdown with their up to 60,000 carbohydrate-degrading enzymes that can be expressed[121,202]. Hence, fiber's real nutritional value for the human body exposes itself after microbial digestion. Unrefined fiber can be regarded as an "indirect" but essential nutrient with important functions: it can be considered a sustained release delivery system, a "vehicle", for carrying nutrients (mostly antioxidants) into the digestive system as they are released by the microbiota during the enzymatic breakdown[201].

Upon digestion by the gut microbiota, the fiber is turned into a variety of short chain fatty acids (SCFAs) that have been shown to be essential molecular drivers regulating human development and the regulation of a balanced immune response as well as human metabolism. "While fiber isn't new or sexy, it may have more evidence backing its benefits than any other single component of the human diet" states an article in the science and health magazine *Elemental*[203].

An increased fiber intake of at least 40 grams per day feeds the microbiota and was shown to improve its richness and stability in healthy individuals compared to a low-fiber diet including merely 10 grams[204]. Increasing the amount of dietary fiber also resulted in more SCFA metabolites in the gut, strengthened the integrity of the intestinal barrier and reduced systemic inflammation in the body and hypertension in obese, diabetics and in patients suffering from inflammatory bowel disease ulcerative colitis[205-209].

Apart from nourishing the microbiota, dietary fiber provides structure. Not only to the plant foods itself, since fiber is the major structural element of plant cell walls, but also to human feces. It bulks up the digestive pulp which gives mechanical impulses to the enteral wall including its smooth muscles and the enteric nervous system, thereby digestive contractions and transit time for better digestion. Every gram of dietary fiber from whole grains, vegetables and fruits improves the intestinal transit time by 30 minutes as a meta-analysis concluded[210].

"Eating enough fiber is Mother Nature's 'cleanse' because
it helps the body eliminate waste products from the
gastrointestinal tract" as nutritional experts point out in an
article about how to communicate the benefits of increased
fiber intake to the American public[211]. In addition to the
benefit of increased SCFA production by the gut microbes,
a fiber provoked regulated digestion and flushing out of
metabolic waste products may be another elemental reason
why fiber intake of at least 35–50 grams per day is associated
with reduced risk for breast cancer, colorectal cancer,
inflammatory bowel disease, cardiovascular disease and all
cancer mortality[212–218]. Ample fiber intake was even observed to
be related to a 50% lower risk for depression in premenopausal
women indicating that metabolites generated from fiber are
involved in the regulation of the nervous system via the gut
brain circuit[219,220].

It is noteworthy that the benefits of fiber seem to be only
conveyed if the unrefined fiber structure is mostly preserved.
Processed fiber in refined foods in contrast have not been found
to purport the same beneficial effects on the microbiota or even
human health as it lacks the structural complexity needed for its
fermentation by the gut microbiota. For instance, healthy subjects
consuming intact legumes and intact cereal grains had more
acidic stools which also contained more short chain fatty acids
than those subjects eating the ground seed diet in one study[201,221].

Although among all fiber containing foods, whole grains are most
consistently associated with reduced incidence of colorectal cancer,
two large prospective US cohort studies did not find any association
for total dietary fiber intake, but when stratifying for different food
sources, a lower risk for colorectal tumors was observed only for
unrefined, intact cereal grains[222].

Despite abundance of data for its health-promoting actions, particularly whole grains but also other fiber-rich foods including legumes, fruits, vegetables and nuts are under-consumed in nearly all countries as the Global Burden of Disease Study uncovered.[15]

The exceptional benefits of fiber for maintaining gut health and preventing chronic diseases were originally investigated during the 1960s and 1970s, and have been prominently promoted since then, by the Irish doctor Dr. Denis Burkitt and his colleagues. During that time, they underwent public health studies in rural Uganda and found that the Africans had a significantly larger intake of dietary fiber relative to Westerners which significantly correlated with a complete lack of chronic, "Western", diseases such as type 2 diabetes, obesity, cardiovascular and digestive diseases and colorectal cancer. Dr. Burkitt, who later became known as the "fiber man", reported that rural Africans passed stools that were up to five times greater by mass, had intestinal transit times that were more than twice as fast and ate three to seven times more dietary fiber (60–140 g versus 20 g) than Western people in Europe and the United States[223].

In 1974, Burkitt and his fellows proposed a theory which became well known as the "fiber hypothesis": "Available evidence suggests that the return of an adequate amount of cereal fiber to the diet would virtually abolish constipation, and almost do away with the need for laxatives. It seems likely that it would also greatly reduce the prevalence of certain common, painful and often fatal diseases that are characteristic of modern Western civilization, and it seems possible that it might significantly reduce the incidence of large intestinal cancer"[224].

However, the molecular mechanisms by which dietary fiber could confer positive health effects were unknown by then.

Burkitt and others speculated that the increased transit time and stool size resulting from the high-fiber Ugandan diet may dilute potentially hazardous microbial-metabolic products and quickly remove them from the colonic lining[225]. Although this was a very valid contemporary theory at that time and partially even holds true today, it is only one part of the story. More than that, current evidence rather suggests the beneficial effects of dietary fiber are primarily attributable to its bacterial fermentation and its conversion into anti-inflammatory, anti-proliferative SCFA molecules which are even shown to counteract the synthesis of toxic metabolites, which we will discuss in more detail soon.

A consistent deprivation of dietary fiber, as usually the case when the regular diet is centered around fiber-free animal based and processed foods rather than unprocessed, plant-based foods leads to a microbial "famine". According to microbiome researchers Erica and Justin Sonnenburg, we are "starving our microbial self" and ultimately depriving the human body of health-maintaining metabolites[226].

In pre clinical animal studies, scientists observed that a chronic lack of dietary fiber reduces the diversity of bacteria in the gut[227]. This effect could not be reversed fully even when fiber was reintroduced, with the effect even compounding over multiple generations.

When gut microbes are starved, some species die off. A number of human trials confirm that a long-term low fiber diet leads to a significant reduction in microbial diversity, diminished SCFA production and results in a profound microbial dysbiosis[228–230]. The drastic microbial composition shift destabilizes and disrupts the overall microbial ecosystem stability. The bacteria begin to switch to an alternative energy source and become "cannibalistic": they start feeding on their own habitat, the colonic mucus layer that protects the gut cell wall from invading pathogens and food particles, which becomes constantly further degraded. The erosion of the colonic mucus layer leads to a diminished protection of the gut cell wall[231]. The manifestation of an intestinal barrier defect, also often described as a "leaky gut", is associated with most chronic diseases, an enhanced pathogen susceptibility in individuals and has been proven to precede the onset of an inflammatory flare up in patients with inflammatory bowel disease, indicating a causal relationship[78,232–235].

Critically, the microbiome's low production of health-protecting metabolites is accompanied by a shift in its metabolism and the generation of substances that are documented to act inflammatory and carcinogenic to the host. "Although the changes in diet following the Industrial Revolution provided more food for growth, the diet became progressively depleted in colonic food, namely fiber, leading to a shift in gut microbiota metabolism that does not provide for colonic mucosal health" as Prof. Stephan O'Keefe describes in an article[236].

In fact, there is increasing evidence that a dysbiotic microbiome marked by a low synthesis of healthy metabolites and increased production of harmful metabolites, is the second part of the story: this toxic microbial shift turns out to be an underestimated risk factor for the development and the progression of several chronic diseases and cancer.

Chapter

6

THE TOXIC MICROBIOME

DOI: 10.1201/9781003212447-6

"It's Not the Fiber, It's the Animal Protein"

In 2020, a research report published by the Produce for Better Health (PBH) Foundation revealed that Americans across nearly all age groups eat less than half of the intake of fruit and vegetables recommended by the Dietary Guidelines for Americans (DGA). Precariously, since the beginning of monitoring in 2004 unprocessed plant food consumption has constantly been declining by 10% with an even steeper drop in the last 5 years[237,238].

"It is no exaggeration that we are in the midst of a fruit and vegetable consumption crisis in our country. Further, this underconsumption is not only pervasive among all age groups but it is also persistent", comments Wendy Reinhardt Kapsak, MS, RDN, President and CEO of PBH. "A decline in fruit and vegetable eating occasions does not bode well for the future of fruit and vegetable intake and, most importantly, Americans' health and happiness"[238].

While Americans chronically and consistently underconsume healthy, fresh produce every day, they upscale on meat, eggs, refined grains and processed foods instead. Throughout the last two decades, meat intake alone has been increasing constantly and now exceeds the DGA recommended tolerable daily serving of around 3.3 oz nearly by the order of 4[23]. According to the *National Health and Nutrition Examination Survey (NHANES)*, nearly 70% of Americans above age 1 eat a multiple of the official daily recommendations for poultry and eggs and more than 90% overconsume refined grains while they neglect whole grains[239].

Although 67% of Americans think they consume enough fiber, the NHANES survey data prove otherwise: 95% of Americans are not getting enough of fiber-rich plant foods, not to mention unprocessed fiber[211,239]. Therefore, the US Dietary Guidelines for Americans even declared fiber a "nutrient of concern for the general US population"[240].

Dietary fiber from plant foods is consistently shown to be one of the most powerful prevention measures against the development of chronic diseases and cancer. People can reduce their individual risk for colorectal cancer by at least 1% for every gram of dietary fiber as a number of prospective cohort studies and meta-analyses conclude[41,216,241]. Although this does not seem much, the lack of fiber-rich plant-based foods poses one of the biggest health risks in our societies and health care systems. The Global Burden of Disease (GBD) Study found that low intake of fiber-dense whole grains alone (below 125 g per day), was the leading dietary risk factor for death and disease in the United States and Germany[15]. Another study funded by the World Cancer Research Fund (WCRF) published in 2021, confirmed that insufficient intake of dietary fiber is the highest lifestyle risk factor for colorectal cancer in Western nations, accounting for around 30% of the cases. According to the WCRF more than 60% of all diagnoses would be preventable by meeting simple lifestyle measures: sufficient dietary fiber, maintenance of a healthy Body Mass Index (BMI) below 30, less alcohol, less meat and regular physical activity of around 3–5 hours per week[42]. However, the worrisome fiber gap in Western populations only tells half the story about the increased risks of chronic disease and colorectal cancer.

In the 1970s, the "fiber man" Dr. Denis Burkitt claimed that the rarity of digestive diseases and colorectal cancer in black Africans as opposed to the modern African Americans and Western nations can be attributed to the high-fiber content of the traditional African diet. However, in the 1990s, Dr. Burkitt's "fiber hypothesis" has been called into question by studies showing that the modern

African has meanwhile adopted a low-fiber content in their diet, yet colon cancer remained rare anyways[242,243]. At that time the African colon cancer incidence rates were still 17 times lower than in high income America or Europe.

What Is behind the African Paradox?

"The low prevalence of colon cancer in black Africans cannot be explained by dietary 'protective' factors, such as, fiber, calcium, vitamins A, C and folic acid, but may be influenced by the absence of 'aggressive' factors, such as excess animal protein and fat, and by differences in colonic bacterial fermentation" as Dr Stephen O'Keefe explains, who was involved in studying the African paradox. "Colorectal cancer remains rare in semi-urbanized agricultural-commercial communities in Africa, who continue to consume >50 g of fiber per day, indicating that we do not have to return to hunter-gatherer status to avoid colorectal cancer, we just need to pay more attention to the nutritional needs of the colon"[243]. Intestinal epithelial cell proliferation was significantly lower in rural and urban blacks than whites while healthy gut microbial fiber fermentation was two to three times higher. The researchers proposed that the increased risk associated with excessive consumption of "aggressive factors" had been annulled by components of protective foods including fiber and folic acid. They concluded: "The rarity of colon cancer in Africans is associated with low animal product consumption, not fiber"[243].

Since Dr. Burkitt's Ugandan studies in the 1970s, Africans now are about to catch up on animal product consumption from formerly 25 g to currently 98 g per day which correlates with the rise of colorectal cancer incidence in the last years[244]. Albeit still 5 to 10 times lower, the cancer incidence gap between Africans and Americans is already shrinking[32,244].

In fact, apart from the dietary fiber gap in Western countries the increasingly high consumption of meat, eggs and dairy turns out to be the other, less told, half of the story about the root causes for the current epidemic of chronic diseases.

Eating less disease-protecting fiber-rich plant foods while *simultaneously* consuming high amounts of low-fiber animal foods synergizes and seems to amplify the risk for developing cancer and chronic diseases.

Depending on the study, specifically processed and red meat intake may increase colorectal cancer risk by up to three-fold in people consuming more than 120 g of meat every day[41,245].

The International Agency for Research on Cancer (IARC) and the World Health Organization have classified meat as carcinogenic[246,247]. Consumption of processed meat has been classified as *carcinogenic* (group 1), while consumption of red meat is labeled as *probably carcinogenic* to humans (group 2A).

Although the association between a low-fiber Western diet and colon cancer as repeatedly documented in epidemiological studies does not prove causality, there is already a body of strong and convincing experimental and clinical evidence from interventional studies that "specific components of animal products" dramatically affect the microbial metabolism, increase the intestinal epithelial proliferation, provoke a systemic immune system response and thereby induce tumorigenesis in humans. Researchers stress that "possibly the strongest evidence is against specific components of animal products"[243].

Which "specific components" in animal foods, however, induce the microbiome metabolism to become toxic and why or how do they influence inflammation and colorectal cancer development?

PROTEIN FERMENTATION

In 1904, the American pathologist Dr. Theobald Smith paved the way for today's well-needed paradigm shift in microbiome science: "It is what bacteria do rather than what they are"[248].

Soon after in 1909, microbiologist Arthur Isaac Kendall published an article in the *Journal of Biological Chemistry* in which he offered visionary thoughts about many of today's hot microbiome research questions and challenges. Like Dr. Smith, he challenged the narrow "bacteriocentric" point of view among scientists at that time and suggested a more progressive idea: "It is becoming more and more evident that the problem of intestinal bacteriology must be approached from the dynamical rather than from the cultural standpoint"[249].

Arthur I. Kendall was one of the first researchers who studied how foods can modify the microbial metabolism. Kendall observed in simple experiments that a protein rich diet significantly alters the composition and biochemistry of the gut inhabitants. He isolated the fecal flora from protein fed animals and grew them in media "of the same fundamental composition as that of the diet which originally nourished them"[249]. Kendall detected unusually "large amounts of gas while there was a gradual diminution in the acidophilic flora grown in acid broth tubes. Coincidentally, products of the decomposition of protein began to appear in the urine"[249].

Ten percent of ingested, generous protein meals escaped human digestion and made their way down to the colon where it was metabolized by the gut microbiota, a process called "putrefaction"[250]. Proteins that reach the colon are fermented anaerobically and decomposed into a range of metabolites, including branched-chain fatty acids such as isobutyrate, isovalerate and isocaproate, as well as indoles, phenolic compounds, sulfides, ammonium, histamine and oxaloacetate[251]. In contrast to fiber, however, protein is not the preferred major substrate for the microbiota. Metabolites and gases resulting from the microbial protein fermentation are shown to be deleterious for the intestinal epithelium when exceeding a certain threshold. "These putrefactive products are generally considered to be toxic and to cause adverse effects on the colonic epithelium so proteins are not considered viable as prebiotics" as a group of Polish researchers raises concerns[252].

It may be important to note that not all amino acids of ingested protein sources are fermented into toxic products by gut microbes. It is not so much the protein catabolism per se that may impact the host negatively, but instead the degradation of specific sulfur-containing amino acids (SAA), methionine and cysteine, in addition to an overall increased protein fermentation activity in response to an excessive protein intake, which depicts the standard diet pattern in the United States and many western European countries[251].

Importantly, even sulfur containing protein is not inherently toxic to the microbiome or the intestinal epithelium. SAA *can* be toxic when surpassing a certain threshold in the gut yet in low concentrations they are deemed essential for the human body. Hence, SAA are considered to be a double-edged sword. Researchers describe H_2S even as a "Janus-faced metabolic product"[253].

How Much of Ingested Sulfur Containing Protein Sources Is Too Much?

SAA are highly abundant in all animal tissues with significantly lower amounts per unit in protein rich plant-based foods. The SAA content in legumes and some nuts for example is ~20–25% of that in animal protein[254,255]. Other plant foods often even contain only 10% of that in animal protein. At low concentrations, H_2S can play an important role in inflammation resolution in colonocytes, and in turn, may support tissue repair, whereas excessive gut luminal concentrations interfere with colonocyte energy metabolism[256-258]. Regular portions of plant-based sources of dietary SAA, however, are deemed sufficient to easily meet the daily sulfur requirements of 15 mg per kg of body weight per day[259].

In contrast, people consuming predominantly animal protein-based diets are shown to synthesize high amounts of the sulfur protein degradation products in their guts, including the "rotten egg gas" hydrogen sulfide (H_2S).

For example, an animal protein-based "Low-Carb diet", which is inherently low in fibrous carbohydrates, changed the bacterial ecosystem in people's guts resulting in an overabundance of bile-tolerant and putrefactive microorganisms with increased capacity to produce protein fermentation-derived metabolites[260]. Several controlled feeding trials demonstrate how switching people from a fiber-rich diet (with little animal protein) to a low-fiber diet (with high amounts of animal protein) resulted in an overgrowth of putrefactive and sulfur metabolizing microorganisms in their gut within a couple of weeks which resulted in significant amounts of H_2S but only a low concentration of short chain fatty acids (SCFAs)[34,198,261]. In response to the changes in microbiome metabolism, colorectal cancer and epithelial proliferation biomarkers markedly increased in the subjects' intestinal tissue biopsies, which was accompanied by intensified immune cell infiltration and immune cell activation in the gut tissue in comparison to the samples from individuals following a diet high in fiber and low in animal protein[34].

H_2S and an overrepresentation of sulfidogenic bacteria in the gut have been identified as a potential environmental risk factor contributing to human colorectal cancer development and inflammatory bowel disease.

In experimental studies, researchers observe that H_2S disrupts the intestinal mucus barrier, acts proinflammatory, impairs cytochrome oxidase, induces DNA damage in intestinal epithelial cells and promoted colonic tumor growth[251,262-266].

Indeed, the metabolic transition of the microbiome from its default production of anti-inflammatory SCFAs towards the predominant synthesis of toxic metabolites, resulting from the microbial digestion of animal products and mucus, has been witnessed in samples from colorectal cancer patients but not in healthy people[85,86,101,267].

Scientists observe a very similar pattern in the guts of patients with the inflammatory bowel diseases (IBD). A meta-analysis of 160 studies revealed that a significant abundance of sulfate reducing bacteria generating three to four times elevated toxic H_2S concentrations was present in IBD patients' guts compared to healthy controls[268]. Furthermore, substantial evidence exists proving the proinflammatory role of H_2S in IBD development[263,269,270].

The microbial dysbiosis, marked by an abundance of protein-fermenting microbes and toxic metabolites, present in colorectal cancer and IBD patients probably comes as no surprise as the consumption of meat and animal product-based diets are strongly associated with the development of these conditions[271].

IBD patients consuming more meat have an up to three-fold increased relapse risk, while conversely removing foods such as milk, cheese, meat and eggs improves the symptoms in IBD patients significantly and was shown to prevent a flare up of inflammation in their guts for up to 5 years[272,273].

Apart from colorectal cancer and inflammatory bowel disease, animal protein-degrading bacteria and their increased protein fermentation ability also seem to play a significant role in modulating visceral pain in inflammatory bowel syndrome and even for the development of autism[274-277].

Intriguingly, there was found to be an antidote to the toxic activity of the Janus-faced sulfur substance: dietary fiber.

In fact, the metabolic process of SCFA generation on the one hand and SAA degradation on the other hand compete. In particular, butyrate and H_2S are fighting a constant battle in the gut lumen. Healthy intestinal epithelial cells depend on the availability of short chain fatty acids such as butyrate as a major food source. The colonocytes oxidize butyrate via the enzyme acyl-CoA dehydrogenase. However, the enzyme is increasingly inhibited the more H_2S is present[278]. Conversely, SCFAs block the synthesis of hydrogen sulfites and other toxic protein metabolism by-products such as ammonia, p-cresol and phenols by competing for enzyme substrates and impeding the enzyme activity[278-280].

Abundant dietary fiber intake therefore may counteract at least some of the negative downstream effects of animal protein metabolism in the gut also corroborating the African paradox because gut diseases and cancer are still so rare in Africans despite their raising animal protein consumption.

In stark contrast, the average adult American rather emphasizes a "high-protein/low-carb/low-fiber" diet pattern and therefore exceeds the Estimated Average Requirement (EAR) for SAA by far. In 2020, an American study analyzed data from more than 11,000 Americans and found the mean SAA consumption was indeed > 2.5-fold higher than the EAR[281].

Independent of the protein intake, an excessive intake of SAA was not only shown to be associated with several gastrointestinal disorders, colorectal cancer or autism, but also with a significantly elevated cardiometabolic disease risk[281]. Researchers explain that "low SAA dietary patterns rely on plant-derived protein sources over meat-derived foods" and suggest "given the high intake of SAA among most adults, our findings may have important public health implications for chronic disease prevention"[281]. Indeed, in addition to a lower risk of colorectal cancer, inflammatory gastrointestinal diseases and heart disease, a dietary restriction of the SAA, methionine and cysteine, was found to cease tumor growth, sensitize colorectal cancers to chemotherapy in humans and significantly extended the lifespan in experimental animal studies[282–284,259].

In conclusion, consuming less animal protein while simultaneously increasing plant foods intake may be a simple yet powerful approach for the prevention of a toxic microbial metabolic shift associated with gastrointestinal and cardiovascular diseases, autism and several types of cancer.

FAT TOXICITY

High animal product consumption does not only come with the side effects of the microbial protein catabolism but is unequivocally accompanied by saturated animal fat intake and its consequences.

Growing evidence associates a high-fat diet, especially its most extreme forms "ketogenic diet", "Atkins" and "Paleo", with a number of human cancers, including breast, ovarian, hepatocellular, pancreatic or colorectal cancer and also with an increased risk of developing chronic diseases such as inflammatory bowel disease, obesity and heart disease[271,285,286].

The common denominator between high-fat, high-protein and low-fiber diet patterns and disease development once again was found to be a toxic microbiome.

According to experimental and human studies, a high-fat Standard American Diet (SAD) which is usually marked by low-fiber content and centered around animal products, disrupts the microbial ecosystem in the gut resulting in a diminished protective mucus layer and increased intestinal permeability: A prerequisite for the entry of toxic metabolites generated by a dysbiotic microbiome.

Even a short period of days or a few weeks on an animal product-based diet was sufficient to rapidly increase the abundance of fat-, protein- and mucus-degrading microorganisms in the guts of people[198]. This dysbiotic transition is accompanied by a rapid leak in the intestinal barrier that leads to the passage of luminal gut bacteria produced endotoxins, such as lipopolysaccharide (LPS), into the blood. Elevated LPS concentrations in the blood are associated with inflammatory processes in several organs which have been demonstrated to be an initiating event in the development of cardiometabolic diseases such as metabolic syndrome, obesity, non-alcoholic fatty liver disease, heart attack and type 2 diabetes[287].

Humans with obesity and type 2 diabetes have significantly higher plasma LPS levels after a high-fat meal promoting the inflammatory cascades in their adipose tissue compared with lean people[288]. A randomized controlled-feeding trial demonstrated that a 6-month high-fat diet led to high concentrations of toxic microbial metabolites including p-cresol and indole in the guts of lean participants and a reduced synthesis of intestinal barrier-protecting SCFAs by the microbiota[289]. The high-fat group also had an LPS-induced endotoxemia with high plasma inflammatory markers, while a low-fat diet had the opposite effect in people. The researchers concluded that "a lower-fat, higher-carbohydrate diet is likely to be associated with a lower risk of excessive weight gain and increase in waist circumference and a more favorable lipid profile than a higher-fat, lower-carbohydrate diet, which might confer adverse consequences for long-term health outcomes"[289].

In fact, a systemic microbiome-induced endotoxemia turned out to be even more severe after a high-fat meal than after smoking some cigarettes in one study[290]. The LPS circulating in the blood led to a strong endothelial activation in the vascular system of the study participants. Endothelial dysfunction has been identified as a hallmark of the majority of cardiovascular diseases (CVDs), so the researchers suggested fat-induced endotoxemia be considered a novel risk factor for atherosclerosis.

One of the many other features of a high-fat diet-induced microbial dysbiosis is the production of the enzyme β-glucuronidase by select opportunistic microbes. The enzyme plays a key role in xenobiotic-induced toxicity from ingested drugs or food additives[291].

Xenobiotics are substances not normally present in the human organism. For detoxification the orally consumed xenobiotics are transported to the liver, where enzymes inactivate them and add tags such as glucuronic acid to mark the molecules for excretion. Then they are transported to the intestines via bile acid (BA). The chemically modified substances are thereby inactivated and become unusable for human cells. When gut bacteria, however, ramp up the expression of the enzyme β-glucuronidases, the glucuronic acid tag is removed from the drug molecule causing unanticipated side effects in the gut[406]. The anticancer drug irinotecan, for example, has been shown to cause intense diarrhea as a result of increased microbial β-glucuronidase activity[292].

β-glucuronidase activity has also been associated with several human cancers including breast cancer, pancreatic cancer and colorectal cancer, therefore bearing a likely high risk for unpredictable toxicity from their medications in addition to the already documented severe side effects from these drugs, e.g., chemotherapy or other cancer-directed drugs[293-295].

Preclinical evidence suggests that inhibiting the microbial β-glucuronidase enzyme in colorectal cancer patients may be a feasible strategy for improved chemotherapy efficacy and provide a higher quality of life due to reduced toxicity experienced from the drugs[296].

Yet, the most side effect-free approach to lower the enzyme activity naturally is adopting a fiber-rich, plant-based diet low in meat and saturated fat.

A plant-based diet was shown to lower an elevated enzyme expression, at least in seemingly healthy people[291,297].

Secondary Bile Acids

Our daily food choices affect the metabolism of our gut microbes and our long-term health more than any other lifestyle factor.

In an article published in *Science* in 2021, researchers tried to create more awareness for one of the most striking symbols of diet-related chronic diseases being a subtle but serious public health crisis: obesity. "Before COVID, obesity and metabolic syndrome were considered the pandemic of the 21st century. Right now, roughly 40% of the U.S. population is obese, and that percentage is predicted to climb", as one of the authors warned[298].

The study authors observe that a high-fat diet is related to microbiome dysbiosis, can damage intestinal epithelial cells and result in a systemic inflammation; the same profile that is prominently found in people with obesity, diabetes and CVD[285]. "It was known that exposure to a high-fat diet causes dysbiosis – an imbalance in the microbiota favoring harmful microbes, but we didn't know why or how this was happening", as study author Dr. Byndloss said:

> Our research has revealed a previously unexplored mechanism for how diet and obesity can increase risk of cardiovascular disease – by affecting the relationship between our intestines and the microbes that live in our gut. We show one way that diet directly affects the host and promotes the growth of bad microbes.[298]

A high-fat diet-induced microbial dysbiosis is not only directly related to obesity but seems to be a prerequisite for other chronic diseases as well, e.g., colorectal cancer.

In order to digest the dietary animal fat, the liver produces excessive BAs that are secreted into the gut. This was shown to lead to a dysbiotic overgrowth of bile tolerant microbes who generate substantial amounts of pro-inflammatory secondary BA: a known risk factor for colorectal cancer development in humans[299,300].

BA are normal metabolites in the intestinal lumen, that are needed for the digestion of fat and the absorption of lipids, as well as the uptake of cholesterol and fat-soluble vitamins[301]. In addition, BA regulates intestinal epithelial cell homeostasis and balances the immune response in the gastrointestinal tract[302].

However, just like an excessive intake in SAA contained in animal protein-based foods, an excessive ingestion of fat tips the balance for the physiologic concentration of BA from beneficial to toxic. While low, physiologic levels of BA are necessary for a proper digestive process, excessive concentrations of BA can be toxic when fermented by the gut microbiota to secondary BA.

For example, a high-fat diet increases the colonic concentration of BAs significantly compared with low- or normal fat diets[301,303]. After the consumption of a high-fat meal, bile acid concentrations can easily reach 1 mM in the colon[304]. Up to 10% of excessive BA is not reabsorbed into the blood during digestion, hence they serve as substrates for microbial metabolism and undergo biotransformation to secondary BAs, mainly deoxycholic acid (DCA) and lithocholate (LCA), which are associated with a number of chronic inflammatory diseases in humans, and cancer, particularly human colon carcinogenesis[293,301,305-307]. In fact, primary BA converted by the gut microbiota to secondary BA was first proposed as a potential tumor-promoting agent as early as 1939[308].

The microbial modification of excessive primary BA produced by the liver into secondary BA in the colon can cause genomic instability and DNA damage, destruction of the intestinal epithelium layer with increased intestinal permeability[309], triggers inflammation and hence induces the risk of cancer in humans in the colon[310,311,300] as well as in peripheral organs such as the liver[312-314] and the breast[315] when entering the circulation.

Since 1993, epidemiologic studies have shown that subjects who consume a high-fat, meat-rich diet produce elevated levels of fecal and serum secondary BA in similar concentrations as usually detected in colon cancer patients[316-319].

The generation of dietary fat-induced secondary BA as a risk factor for colorectal cancer development was later supported in a prominent cross over intervention study. Rural Africans were fed a high-fat/high-protein/low-fiber (Western-style, SAD) for 2 weeks while a group of Western-lifestyle pursuing African Americans were fed a high-fiber/low-animal traditional African-style diet. This resulted in a massive functional shift of the participants' microbiomes. On switching to a high-fiber/low-fat diet, the colonic butyrate increased by 2.5-fold and secondary BA lithocholate by 70% in African Americans. On the other hand, switching to a low-fiber/high-fat diet in native Africans resulted in a reduction of colonic butyrate content by 50% and an increase in secondary BA by 400%[34]. The colonic mucosa biopsies of the study subjects showed that several colorectal cancer risk markers (Ki67, CD3[+], CD68[+]) increased markedly under the high-fat diet regimen while the markers were reduced in individuals under a high-fiber/low-fat diet.

> These findings are exciting as they show not only that colonic microbial metabolism responds rapidly to dietary modification, but also that these microbial metabolic changes are accompanied by changes in mucosa known to increase or decrease the susceptibility to neoplastic change and cancer risk within 2 weeks

as the researchers point out[34].

Current metagenomic and metabolomic studies report that colorectal cancer patients display a characteristic gut microbial cancer signature, marked by potent protein and fat degradation abilities and high levels of toxic metabolites, further underscoring the significant role animal fat and protein-rich diets can play in colorectal cancer development[101,293].

The effects of intestinal secondary BA on colon cancer development may be obvious, but its entry into the body's circulation makes it an unpredictable risk factor for other cancers, too.

Almost two decades ago, researchers had already identified a role of gut microbially produced secondary BA in human breast cancer development. Researchers found the toxic metabolites in cystic breast tissue in humans[320,321]. Later, studies confirmed that the plasma DCA concentration was 52% higher in patients with breast cancer compared with controls and observed that 82% of breast cancer patients expressed significant amounts of the cellular bile acid receptor[322,315].

Is Dietary Fat Ultimately "Bad" for Us?

According to large epidemiological and intervention studies the composition of dietary fat rather than the total fat content in a meal has the greatest impact on the pathogenesis of colorectal cancer and cardiovascular disease[565]. In particular, trans-, saturated and monounsaturated fats from animal sources are shown to bare a high risk for cancer and heart diseases in comparison to unsaturated fats from plant-based foods such as nuts, which conversely had a strong protective effect.

For example, influential studies, such as PREDIMED and the Health Professionals Follow-Up, demonstrate that adapting a Mediterranean diet based on plant-based fats from nuts and olive oil, is associated with a lower risk for cardiovascular diseases, type 2 diabetes and breast cancer than caloric intake-adjusted low-fat diets or fats from animal products such as meat and fish[323-325]. This can at least partially be explained by a diet-induced change in microbiome metabolism towards increased protein and fat digestion.

The subjects with a low adherence to the Mediterranean diet, those with high red processed meat and fish intake, harbored a gut microbiome with stronger secondary bile acid biosynthesis potential compared to subjects who followed a healthier, plant-based version of the Mediterranean diet[323]. Similarly, results from the prospective Nurses' Health Study report that specifically monounsaturated fatty acids from animal sources were associated with an increased risk of developing colorectal cancer[326].

Although a high-fat content in our diet can manipulate the gut microbiota to behave in a toxic manner, instead of curbing fat altogether. Even more important seems to be the ability to distinguish between animal and plant foods as fat sources, the latter including high amounts of dietary fiber as an additional protective ingredient.

HEME IRON

Increasingly, evidence suggests, there is not a single exact mechanism by which animal protein intake, particularly red and processed meat, is related to cancer. Instead, research suggests a plethora of synergizing factors that increase the risk of cancer and chronic inflammatory diseases.

Apart from endogenous toxic protein and fat metabolites, other meat-derived substances, especially N-nitroso compounds, aldehydes and heme iron are likewise shown to cause DNA damage and are suspected of contributing massively to inflammation and carcinogenesis in human colons.

In recent years, in particular the dietary uptake of heme iron has attracted great attention as a major component of meat to play a role in the etiology of colorectal cancer[327,328]. Iron is an essential mineral and serves as a key component of oxygen carrier molecules hemoglobin in our erythrocytes, the red blood cells, but also of myoglobin in muscle cells as indicated by the red color. Therefore, a certain amount of dietary iron is recommended for optimal oxygen transport to the organs. However, dietary iron can either be obtained from heme iron, present in animal tissues, or from non-heme iron, present in plant-based foods such as whole grains, legumes, nuts and dark leafy greens.

Research has linked heme iron, but not non-heme iron, with an elevated cancer risk. However, the evidence is still inconsistent about a direct involvement of the heme itself. The underlying mechanisms are not fully understood yet, but heme iron has been shown to trigger intestinal epithelial cell hyperproliferation, gut microbiota dysbiosis and DNA damage in preclinical and human studies[262,329–331].

The heme content in red and processed meat products (bacon, hot dog, roast beef, steak, ham, sausage pork) can be ten-fold higher than in white meat such as chicken breast and fish[332,333]. Red meat consumption has been classified as "possibly carcinogenic" to humans by the International Agency for Research on Cancer (IARC), the high-heme iron content being one of the reasons they named[334]. Although several large prospective cohort studies were able to demonstrate that a high intake of heme iron correlates with a higher risk of colorectal cancer, some analyses failed to establish the link, suggesting a so far unknown mechanism likely involving other factors[335–338]. Interestingly, the Cancer Prevention Study-II, *Nutrition* detected a 29% increased risk of colorectal cancer with elevated blood bilirubin levels in patients, a degradation product of heme, implicating an indirect involvement of heme in cancer development[339].

Currently, the evidence points to heme as an indicator for red meat intake and may rather be considered an indirect risk factor for cancer development. For instance, heme iron is a major catalyst in the formation of endogenous N-nitroso (NOC) compounds during protein digestion and facilitates the generation of aldehydes in the gut during fat peroxidation. Both, aldehydes and NOC, build up in the feces of people after eating meat causing severe DNA damage in intestinal epithelial cells, increasing the risk for cancer[340-342]. In fact, the highly mutagenic properties of human feces after meat intake were already described as "fecal water toxicity" in 1977[343].

From the 1960s on, scientists became aware the carcinogenic effects of the nitroso compounds when back then studies associated exposure to NOC with several cancers including brain tumors in children, childhood leukemia, colorectal cancer and gastric cancer[336,344-349].

In a human intervention feeding study, researchers observed chromosome instabilities and micronucleus frequencies, validated markers and predictors of colorectal cancer, as well as cancer-related gene expression patterns indicative for the genotoxic effects of NOCs in colonic biopsies of subjects after red meat intake[350,351].

The accumulating data of concern relating NOC to cancer has led the IARC to classify the toxic substances as a 2A carcinogen: probably carcinogenetic to humans[352,353].

N-nitroso compounds are generated in the intestinal tract via several mechanisms, one of them requires a nitrosating agent (nitrate or nitrite ingested with foods) and protein. The NOC formation can be inhibited though by dietary antioxidants, e.g., vitamin C, occurring in abundance in plant-based foods, which is why nitrate in some plant foods does not result in toxic NOC generation, also indicated by the significantly low cancer rates in people eating plant-based diets[354].

For example, the Shanghai Women's Healthy Study, a large cohort study, investigated the association between dietary nitrate and nitrite intake and the risk of colorectal cancer. While the overall nitrate intake was not associated with colorectal cancer risk, the risk increased more than two-fold with higher nitrate intake among those women with a low vitamin C intake and hence a higher formation of NOC in their guts[355].

The formation of toxic NOCs seems to be exclusive to red and processed meat ingestion, but not to other animal sources such as white meat or non-heme iron from plant-based food sources indicating the catalytic role of heme iron in the process.

Human volunteers given a high red meat diet excreted much more NOC in their stools than controls given no or little red meat, or only white meat[342,356]. In comparison to a protein-rich plant-based diet, a heme-rich meat based diet significantly increased the fecal NOCs levels in people, corroborating the theory of heme as a catalyst for accelerating endogenous NOC formation in the gut lumen[357].

Conclusively, the culprit for accumulating the carcinogenic N-nitroso compounds seem to be the heme from meat consumption but not a high protein intake per se.

Although N-nitroso compounds can also be ingested with foods, primarily from processed meats such as hot dogs and sausages, but also with cigarette smoke, most of them form endogenously in the gut during digestion.

Once more, the gut microbiota plays a significant role in this mechanism.

About 20 years ago, early experimental studies found that endogenous formation of nitroso compounds in the gut required in the presence of an intestinal microflora since researchers could not detect the toxic substances in germ-free animals[358].

A dysbiotic gut microbiome seems to facilitate the synthesis, as the population of NOC-producing bacteria in the guts of healthy people who consumed very little meat were found to be low and the amount of nitroso compounds was neglectable[359,360]. Conversely, excessive dietary nitrite intake fostered the growth of these bacteria and also enhanced lipidperoxidation reactions in the guts[359,360]. Importantly, these processes could be reversed by wiping out the gut flora with antibiotics as demonstrated in experimental studies, further confirming the role of the microbiome in the generation of NOC[361].

Apart from promoting nitroso compounds and fat peroxidation, heme also provokes oxidative stress when gut bacteria metabolize the iron. In a metaproteomics study, researchers found a significant increase in microbial proteins related to iron metabolism and oxidative cell stress in samples from colorectal cancer patients but not in healthy people, which is known to fuel inflammation and cancer development.[185]

Hence, heme iron appears to have a complex function in the process of cancer development which needs to be further elucidated.

Although there is still no direct human evidence for heme iron to be a single causal agent for gut inflammation and colorectal cancer development, current observations strongly suggest heme to act as a strong catalyst in shaping a toxic gut environment.

In consequence, heme turns out to be another one of the multiple synergizing factors on the list relating cancer risk and meat intake with the gut microbiome.

TMAO: THE WAY TO A MAN'S HEART IS THROUGH HIS GUT MICROBIOTA

In 2011, *Nature* published a fascinating News & Comments article titled "The Diet-Microbe Morbid Union" in which the authors stated: "A common dietary component that some people even take as a supplement is converted by the gut microbiota to harmful metabolites linked to heart disease"[362].

The article mentions one of the first solid pieces of evidence for a unique microbiome-derived metabolite to be involved as an independent risk factor in human disease development: Trimethylamine N-oxide (TMAO).

Researchers were screening the blood from patients that had experienced a heart attack or stroke for metabolites associated with the disease and compared the results with those from the blood of healthy people. They found differences in the abundance of three specific metabolites, choline, betaine and TMAO, that were significantly elevated in the patients. These metabolites are resulting from the metabolism of frequently consumed dietary lecithin and L-carnitine intake. The primary sources of dietary choline, lecithin and L-carnitine are meat, egg yolks, dairy products and seafood[363-365].

In animal experiments, researchers were able to confirm a direct link between the intake of lecithin and the level of circulating TMAO which resulted in enhanced atherosclerotic plaque development. Atherosclerotic plaques are a known risk factor for heart attack and stroke, therefore the researchers proposed TMAO as a novel risk factor cardiovascular diseases[366].

Also, the more TMAO is detected in people's blood, the higher the risk of death, myocardial infarction, or stroke, which interestingly seems to be independent from other known risk factors such as cholesterol or high blood pressure that clearly add up on top. For example, people with the highest TMAO concentrations had an up to nine times elevated risk for cardiovascular disease compared to people with the lowest concentration[367].

Human intervention studies demonstrated that a precursor of TMAO, trimethylamine (TMA), can only be produced by gut bacteria[368,363]. In fact, the TMAO level in one's blood already increases after two hard-boiled eggs, a rich source of lecithin, while the TMAO synthesis can be impeded again when compromising the microbiota with an antibiotic treatment[368]. Similarly, studies feeding people isotope-labeled L-carnitine, another substrate for the gut microbiota to facilitate TMAO development within hours, clearly identified the gut microbiota as the critical actor in the production of the toxic molecule[368,364,369].

The gut microbiota metabolizes the dietary lecithin or L-carnitine, respectively, into trimethylamine (TMA) – a gas that smells like rotten fish. Then TMA is further oxidized to TMAO in the liver before it finally enters the bloodstream, spreads to peripheral organs and is finally excreted through the kidneys[362,370].

The microbiome's ability for TMA synthesis is dependent on the amount and frequency of ingested dietary sources of lecithin and L-carnitine, which are animal-based protein sources, respectively. Accordingly, fasting baseline TMAO levels are undetectable among long-term vegan and vegetarian people compared to omnivores. In fact, a vegan's microbiota lacks the ability to produce TMA at all[364].

Yet, the microbiome is a vivid and plastic organ adapting its metabolism effectively to new substrates.

For example, feeding vegans an L-carnitine supplement equal to the amount found in a steak for 3 months, trained their microbiome to adapt its metabolism to produce TMA: after the first month, TMAO plasma concentration was already elevated in the vegan subjects and further increased markedly during the 3-month study period[369].

In contrast, the microbiota of people who frequently consumed foods that are rich sources for substrates for TMAO synthesis but were low in fiber, shifted their metabolism accordingly from default fiber degradation to TMA, protein and fat catabolism.

For example, people following a Paleo diet excluding whole grains and legumes were shown to produce significantly more TMAO in their bodies than people whose diet is tailored according to national dietary guidelines which recommend including whole grains[371].

The study authors concluded: "although the Paleo diet is promoted for improved gut health, results indicate long-term adherence is associated with increased TMAO. A variety of fiber components, including whole grain sources may be required to maintain gut and cardiovascular health"[372].

Even more extreme, a diet comparison reveals that a very low-carbohydrate diet (Atkins or Ketogenic) may contain around 2.5 times more choline than a low-fat, plant-based diet[362]. "It is thus tempting to speculate that a very low-fat diet may reduce the risk of heart disease in part because of its low choline content", as the researchers speculate[362].

The lack of dietary risk factors such as choline or L-carnitine and the high-fiber content in their diet are another piece of evidence explaining why vegetarians and vegans are shown to have a 23% and 42% lower risk of dying from CVD and a lower risk of all-cause mortality, respectively, than omnivores[373].

During the early days of the discovery of the "morbid union" between the microbial metabolites and diet, a lingering question was whether an increased TMAO level in the blood of people contributes to CVD or is simply a marker of disease risk. Since then, countless evidence has accumulated confirming that "the way to a man's heart is through his gut microbiota"[374].

Multiple meta-analyses have independently confirmed that plasma TMAO levels are associated with cardiovascular events and mortality risks across multiple patient populations and geographic areas[365,375,376].

One meta-analysis reports that each 10-μM increase in TMAO in the blood was associated with an absolute 7.6% increased risk of all-cause mortality[375]. Another meta-analyses of 11 studies reported that higher circulating TMAO is associated with a 23% higher risk of cardiovascular events and a 55% higher risk of all-cause mortality[365]. Indeed, subjects with higher TMAO levels have a 36% higher chance of dying earlier than those with lower plasma levels, which affects particularly subjects with impaired kidney function, as the molecule cannot be cleared appropriately and accumulates in the system[374].

As a matter of fact, a higher TMAO level in the circulation system, besides its precursor choline, has also been found to be a strong risk factor for an adverse clinical outcome in patients with heart failure[368], type 2 diabetes[377], kidney disease[378] and is associated with an increased thrombosis risk[379], as well as colorectal cancer development[380–382].

Despite the stunning body of evidence supporting the association between TMAO and disease development, there are some inconsistencies in studies and remaining questions clearly demonstrating the complex role of the molecule and the many factors on which its deleterious effect on the human organs depends on[383].

For example, a Mediterranean diet is widely accepted to be associated with a lower CVD risk which is usually attributed to a regular seafood consumption. However, seafood is a rich source of TMAO. Calling into question the public narrative of fish as the beneficial ingredient in a Mediterranean diet, an analysis of the influential PREDIMED study revealed that the actual disease-protective benefit of a Mediterranean diet pattern may be attributed to the high fruit and vegetable content rather than the fish[384].

Also, there is still some uncertainty about inter- and intra-individual variations of TMAO levels which can depend on many aspects that seldomly receive attention in studies.

For example, there are stark differences in patient populations due to non-standardized sample types (serum or plasma), besides disparities in the total amount and ingestion frequency of foods containing choline, lecithin, L-carnitine or TMAO itself. Furthermore, the concentration of BA secreted in the gut is mostly left unconsidered which however inherently induces the expression of the TMAO synthesizing enzyme, flavin monooxygenase (FMO3), in the liver. Finally, another confounding factor may be hard to consider and adds on to the inconsistencies: the individual microbiota's metabolic ability to generate TMA at baseline as well as the individual kidney clearance rate[385].

Yet, despite these issues, which require further investigation, there is no doubt among the scientific community about the toxicity of TMAO and its harmful effects on human health.

Several strategies are proposed to lower TMAO concentrations in the body in order to lower the risk of atherosclerosis, CVD and cancer.

While some researchers and especially certain industries support "drugging the microbiome" by using FMO3's enzyme inhibitors or antibiotics[298], other scientists simply recommend "the most obvious is to limit dietary choline intake"[362].

Indeed, swapping animal-based meat with plant-based meat alternatives in healthy adults for 2 months was enough to improve their cardiovascular disease risk profile in the blood including TMAO, suggesting substituting plant proteins for red meat can effectively lower the risk for CVD[386,387].

In fact, eliminating the food sources causing the microbiome in liaison with the liver to synthesize TMAO in the first place may be one of the safest, cheapest and effective ways: "the way to a healthy heart may be through a healthy gut microbiota"[374].

TOXICOMICROBIOMICS

Apart from its capability for endogenous production of toxic metabolites from foods, another metabolic trait of the microbiome has also been widely underestimated so far: the conversion of dietary xenobiotics into substances with unpredictable health risks.

In fact, the microbiome is very potent in transforming artificial substances ingested with food to become either less or even more toxic.

Xenobiotics are chemical substances not normally present in the environment of living organisms. Among the most commonly encountered industrial chemicals are persistent organic pollutants (POPs) including pesticides such as DDT or chlorpyrifos, polychlorinated biphenyls (PCBs) as well as dioxins. Although some of them were banned in the late 1970s, they are persistent, are hardly being degraded and have accumulated in the environment in water and soil already since the 1930s. They are taken up by animals and particularly enrich in their fat tissues. As a consequence, 90% of the human exposure to these chemical substances comes from animal foods: fish, meat, eggs and dairy[388,389].

According to a study by the European Food Safety Authority (EFSA) nearly half of the tested foods in the European Union contain a cocktail of environmental toxins of which some even exceed the acute reference dosis[390]. Several of these substances are collectively known as endocrine-disrupting chemicals (EDCs) since epidemiological studies have shown that exposure to them is associated with adverse health effects, including diabetes and obesity, several types of cancers, and negative effects on the immune, nervous and endocrine systems[391-394]. Exposure to PCBs for example can lead to an impairment in the neurological function and behavior in children and increases the risk of childhood leukemia[395].

The mechanisms underlying the deleterious effects of persistent environmental pollutants and pesticides on human health are poorly understood. Experimental studies, however, indicate that one of them leads through the microbiota metabolism. Many preclinical investigations gave proof, that exposure to environmental chemicals, even at doses that are officially considered to be safe, were observed to affect the microbiota composition in the gut of experimental animals and disrupt its default metabolism[396-399]. Furthermore, this was accompanied by a compromised protective intestinal barrier and a subsequent leakage of the bacteria-derived endotoxin LPS into the body which led to a low-grade systemic inflammation, insulin resistance and obesity. Transplanting this chemically disrupted microbiota into healthy experimental animals resulted in inflammation and insulin resistance, while a treatment with antibiotics reversed the condition within a couple of weeks, suggesting a strong case for the microbiota in the toxicity of xenobiotics[399,400].

To date, 1369 environmental pollutants have been found to be metabolized by the human gut microbiome[401].

Consequently, the microbiota's ability to modify and supposedly even amplify the toxicity of xenobiotics, renders the gut microbiome into an unpredictable factor when assessing the actual effects of drugs or food additives on human health.

Xenobiotics are not only readily ingested with foods or drugs but do also arise from the pyrolysis of protein when cooking and processing meat, including polycyclic aromatic hydrocarbons (PAHs), heterocyclic amines (HCAs) and nitrosamines (NOCs).

Apart from the carcinogenic effects of nitrosamines, as discussed previously, research has accumulated substantial evidence for an increased cancer risk in humans after exposure to HCAs[402]. Worrisome, in recent years more than 25 different HCAs have been identified in regular food products, formed via processing from creatinine, creatine, hexoses, amino acids and some dipeptides present in the muscle of meat and fish[403].

The microbiota's enzyme portfolio can transform xenobiotics into genotoxins or may reverse the detoxification implied by the host metabolism as observed in experimental studies[404,405]. Xenobiotics are often conjugated to glucuronic acid in the liver (one of the pathways of phase II detoxification in the human body), stored in the gallbladder and released into the gut with bile during the digestive process. When the conjugated xenobiotic enters the intestine, the microbial enzyme β-glucuronidase may cleave the molecule and release the unconjugated xenobiotic turning it back into a toxic molecule[404,406].

Since founded research on the microbiota's role in transforming xenobiotics is still scarce and widely unexplored in humans, wider cohort studies are urgently needed to obtain a clear perspective of the long-term impact on human health resulting from the intake of xenobiotics, considering the still increasing consumption of processed and animal foods globally.

A Toxic Microbiome for Diagnostics: Useful for Predicting Chronic Diseases?

The paradigm shift from the former focus on microbial taxa towards the functionality of the microbiome massively facilitates our understanding of its role in human health and disease.

Microbial compositions slightly change every day within each of us and differ even more widely between individuals. While less than 50% of the species-level bacterial taxa are similar among different individuals, more than 80% of the microbiome's functional traits were shown to be shared between people[9,159].

Overly simplified this may mean we can stratify microbiomes by a "healthy" behavior and a "disease-related" functional pattern, as represented by a specific set of microbial metabolites, respectively. It may even be suitable to consider the taxonomic composition as rather secondary.

Metabolomics is the comprehensive study of the metabolome, the repertoire of all small molecules present in cells, tissues and body fluids[558]. At the center of this concept, is the idea that a person's metabolic state provides a close representation of one's overall health status.

The metabolomics approach has the potential to pave the way for tremendous progress in microbiome science with less conflicting and more robust, replicable data due to lower inter-individual variations. Detecting changes in microbial behavior by translating the microbial "language" (their metabolites) prior to chronic disease manifestation can be a powerful opportunity for the development of rapid, reliable and less invasive techniques in clinical diagnostics.

The food metabolome, or "nutriome", is defined as "the part of the human metabolome directly derived from the digestion and biotransformation of foods and their constituents", which can be measured in our blood, saliva or urine[407]. With >25,000 compounds known in various foods, the food metabolome is extremely complex, considering the even larger array of breakdown products since 30–40% of the blood metabolites are derived from or additionally modified by the gut microbiome[9,407,408]. The strong relationship between the long-term healthy or unhealthy diet patterns and a respective, corresponding host-microbial metabolite pattern has already been documented in healthy individuals[409,410].

These data may become clinically relevant for characterizing the shift from a healthy to an unhealthy metabolic phenotype and possibly disease prediction.

Since many studies were able to associate a specific set of microbial genes or metabolites in patients' blood with several diseases, a metabolome analysis may even predict a disease risk well in advance before the clinical manifestation of symptoms.

Prominent metabolite signatures already identified in patients with cardiac events and CVD, chronic kidney diseases, colorectal cancer, type 2 diabetes amongst many other conditions involve fat and protein degradation, choline metabolism and TMAO[86,368,411,412].

Increasing evidence supports the predictive value of this metabolic portfolio. Particularly TMAO seems to crystalize as a unique identifier and significant predictor of a poor prognosis for several chronic diseases in more and more studies.

For example, a high betaine and TMAO plasma level was found to be significantly associated with a three times increased risk of cardiovascular events including heart failure in diabetics[413,414]. Also, in persons with acute coronary syndrome their TMAO level appears to be a strong and independent predictor to suffer from early death[414]. Patients with the highest concentrations had an up to 11 times higher risk of death than those with low TMAO levels.

Apart from that, high TMAO levels is also shown to be an early marker of atherosclerosis and to predict fatal outcomes in chronic obstructive pulmonary disease (COPD), chronic kidney disease and pneumonia patients[415-417,378].

TMAO is not the only critical molecule; other microbial metabolites in the blood such as taurine and betaine were likewise shown to be positively associated with and causally related to an increased risk of inflammatory bowel disease, for instance[418].

Using a combined metagenomics and metabolomics approach researchers are confident to identify specific metabolic signatures that clearly distinguish advanced stages of a disease from early stages, suggesting this to be a fast and non-invasive technique to predict a disease development.

For example, researchers were able to distinguish patients with non-alcoholic fatty liver disease (NAFLD) between severe cases with liver cirrhosis and very early stages of fibrosis only based on their metabolic signatures[419].

Similarly, several studies find a robust functional microbiome signature in colorectal cancer patients' fecal samples that has not been observed in healthy controls. Specifically, the researchers detected a signature for enriched protein, choline and mucin degradation as well as depleted fiber degradation, suggesting that the cancer patients' microbiomes adapted to a fat- and meat-rich diet for a long-time already, confirming these foods as a high-risk factor for fostering colorectal cancer development as discussed in earlier chapters[86,101]. The microbiomes' adaptation to a harmful protein and fat metabolism, indicated by a range of metabolites such as TMAO, sulfur compounds or choline amongst others, is measurable at early stages in colorectal cancer development. Scientists were able to stratify non-neoplastic, adenoma and early- and late-stage colorectal cancer patients based on the metabolic signature in the fecal samples[267,293]. Because the detection of precancerous lesions with high sensitivity is still a big challenge, these promising results may open the exciting possibility for very early cancer detection from fecal microbial markers.

However, larger, independent cohorts, standardized sample preparation and robust metabolomics analysis methods are needed to further evaluate the diagnostic value of metabolites for disease or outcome prediction in clinical settings.

The most effective way to fight cancer though is primary prevention, particularly through a plant-based diet, which prevents the formation of a toxic microbiome and ultimately lowers the risk for chronic diseases and several types of cancer. Public health researchers therefore demand: "primary prevention must be prioritized as an integral part of global cancer control"[420].

Chapter

7

HOW TO BUILD A HEALTHY GUT MICROBIOME AND PREVENT CHRONIC DISEASES

DOI: 10.1201/9781003212447-7

THE "TRUE" HUMAN DIET: ARE HUMANS REALLY OMNIVORES?

The Greek philosopher Pythagoras is supposed to have said "Oh, how wrong it is for flesh to be made from flesh!" Was he right?

Apparently, people have been debating the issue of animal protein as part of a natural human diet for at least 2500 years starting out as a question of the morality of eating other animals while today other issues add up such as questions about health as well as sustainability of animal food production in our global food system. Facing the threats of three crises, the slow global ecosystem breakdown of our natural environment at a large scale, the collective gut microbiome disruption in modern populations on a microscopic scale as well as the public health crisis of diet-related chronic diseases as the leading killer of humanity in the twenty-first century, humanity is urged more than ever to discuss the issue of how to feed the world.

However, the foremost, underlying question to solve the problem is: what is the natural human diet? What are the foods our bodies are designed to digest best that also keep us healthy? Are these foods sustainable enough to feed a growing world population?

A heavily animal product based diet is not only detrimental for our gut microbial health and the global environment, but also it may not even be appropriate for our human biochemistry and anatomy. Theories of human evolution have postulated that the increased amount of meat on our plates permitted the increase in brain size and therefore emergence of modern man. However, comparative studies of primate intestinal tracts, e.g., do not support this hypothesis. "It is likely that, while meat assumed a more important role in hominid diet, it was not responsible for any major evolutionary shift" as the French anthropologist Prof. Claude Marcel Hladik and his colleague Dr. Pasquet conclude based on their extensive research[421]. "Meat has been obviously consumed, but it is unlikely that animal flesh especially lean meat was a staple food for long periods" states Hladik[422].

According to Hladik's and other research works, humans have actually evolved as habitual meat eaters. For example, humans still have the capacity to digest heme iron and other porphyrin-iron compounds derived from meat, whereas strictly herbivorous mammals cannot absorb these compounds. Therefore, humans cannot be classified as herbivores either. Instead, the large body of scientific evidence considers humans as "frugivores", whose body physiology is perfectly adapted to eating fruit-like produce of plants such as nuts and seeds, grains, fruits, roots, vegetables supplemented with young shoots, leaves and at most occasionally animal matter (predominantly invertebrates such as snails, worms or insects) to meet their protein demands[423].

Looking into the evolution of our human physiology, researchers unambiguously agree on which foods the human body, the anatomy of our digestive system and human microbes predominantly prefer for optimal digestion and obtaining nutrients. The same foods happen to come with powerful benefits to effectively protect us against diseases. Ranging from an abundance of nutrients not present in any other food source to metabolites resulting from its digestion such as short chain fatty acids: plants.

Humans seem to have a foot in each clamp: neither carnivore nor herbivore. For example, the appendages of carnivores are claws; those of herbivores are hands or hooves. Body cooling of carnivores is done by panting, herbivores, by sweating. Carnivores drink fluids by lapping; herbivores, by sipping. The teeth of carnivores are sharp; those of herbivores are mainly flat for grinding. The intestinal tract of carnivores is short (3 times body length); that of herbivores, long (12 times body length)[424]. Modern human guts happen to be somewhere in between. Indeed, since humans diversified from our last common ancestor with apes, human colon size has reduced to around 30% of true herbivores, while at the same time still having a massive absorptive area for nutrients[424]. Measurements of human gut size and form by researchers support that human gut anatomy and pattern of digestive kinetics is on the best fit line of the frugivores[422]. The rather ambiguous term "omnivore" for humans is used to either describe the seasonal and geographical food variety humans are adapted to, or to emphasize that a meat supplement is eventually included into a frugivorous diet. "However, one should notice that the largest primate species, especially anthropoids, use mainly vegetable matter to meet their protein requirements. Chimpanzees, may occasionally eat the meat of small mammals, but they do not cover their protein requirements from this source" as Prof. Hladik comments on the debate about the true human diet[425].

Another clear indicator for identifying humans as mostly plant-based eaters is that omnivore humans who regularly eat animal product are at a significantly higher risk to develop atherosclerosis and heart diseases compared to vegetarians. For example, according to a meta-analysis and data from the Adventist Health Studies, vegetarian and vegan males experienced a 23% and 42% risk reduction in dying from cardiovascular disease, respectively[426,427]. Vegan males even had a 55% risk reduction for ischemic heart disease.

Today, atherosclerosis and heart disease are one of the leading killers of omnivore man.

In contrast, atherosclerosis does not affect carnivores. Meat-eating animals like dogs, tigers or cats can load their diets with saturated fat and cholesterol while not developing a single atherosclerotic plaque[428]. Experimental studies demonstrated the only way to get a carnivore to develop atherosclerosis is to remove the thyroid gland[429].

In addition, carnivores can produce their own vitamin C, whereas all other animals and humans depend on adequate amounts of fresh, plant-based foods as an abundant vitamin source on a daily basis. Humans lost the ability to produce vitamin C from glucose at some point in our evolution due to a genetic mutation in an enzyme that facilitates ascorbic acid synthesis from glucose[430–432].

Consequently, although meat eating certainly played some role in hominid history, yet the hominid flexible gut anatomy permitted adaptation to various diets. According to the primate researcher Marcel Hladik:

> there is no evidence to support theories such as a change in gut anatomy that allowed carnivorousness and a simultaneous increase in brain size. Alternatively, the early cooking of gathered foods – and the nutritional, behavioural and social consequences of this pattern – could have been a major milestone in the hominization process[421].

"THE PALEO DIET IS A MYTH"

Neglecting the anthropological and current microbiome research, "cavemen diet" gurus have built a strong case for discordance between what we eat today and what our ancestors evolved to eat. Paleoanthropologists like Prof. Peter Ungar complain about the mislead understanding and modern interpretation of the natural human diet: "The idea is to eat like our Stone Age ancestors – you know, spinach salads with avocado, walnuts, diced turkey, and the like"[433].

Most Paleolithic diet disciples agree on a diet rich in protein and omega-3 fatty acids which they base on grass-fed cow meat and fish accompanied by vegetables and nuts, while sparing out dairy, potatoes and highly refined and processed foods. This is an improvement in dietary quality compared to a standard American diet which relies heavily on processed foods and little vegetables or nuts. However, healthy food groups such as cereal grains and legumes are also out.

According to Prof. Peter Ungar, however, there is nothing new about cereal consumption and gluten is not unnatural either.

> Despite the pervasive call to cut carbs, there is plenty of evidence that cereal grains were staples, at least for some, long before domestication.
>
> Indeed, researchers at the archaeological site *Ohalo II* on the shore of the Sea of Galilee found a collection of more than 90000 plant remains representing cereals (emmer wheat, barley), almonds, raspberries, grapes, wild fig, pistachios and various other fruits and berries. Owing to the excellent preservation, a stunning 142 different species of plants were clearly identified, suggesting that a rich diversity of grains and other fiber sources was consumed by the site inhabitants during the peak of the last ice age, more than 10,000 years, well before these grains were even domesticated[434].
>
> Paleobotanists have even found starch granules trapped in the tartar on 40,000-year-old Neandertal

teeth with the distinctive shapes of barley and other
grains and the characteristic damage to the kernel
structure that comes from cooking[433,435].

According to Paleobiologists, there is no universal Paleolithic diet.

Apart from the fact that a precise reconstruction of nutrient
composition of particular hominin species in evolution is difficult,
and our understanding about it is incomplete, diet composition
has been changing constantly depending on era, geography and
season. "What was the ancestral human diet? The question itself
makes no sense", as Dr. Ungar finds fault with the popular
Paleo diet movement[433]. In a similar manner, Rob Dunn, a biologist
at North Carolina State University, puts the current understanding
of the Paleo diet into question

> Which Paleo diet should we eat? The one from twelve
> thousand years ago? A hundred thousand years ago?
> Forty million years ago? If you want to return to your
> ancestral diet, the one our ancestors ate when most
> of the features of our guts were evolving, you might
> reasonably eat what our ancestors spent the most time
> eating during the largest periods of the evolution of
> our guts, fruits, nuts, and vegetables – especially
> fungus-covered tropical leaves[436].

Dental microwear and stable isotope analyses from early African
hominins dating back 7 million years uncovered they ate up to
80% plant-based foods similar to that of grass-eating warthogs
or zebras[437]. The preponderance of evidence supports the idea
that plants predominated in nearly all hunter-gatherer groups.
Depending on the geographical region the proportion of plant foods
in the diet of hunter-gatherers is estimated by Eaton and Konner

to be up to 80% by weight in inland and semitropical habitats and up to 90% in coastal areas even[438]. The San tribe in Botswana's Central Kalahari took about 70% of their calories from carbohydrate-rich, sugary melons and starchy roots; and even Neanderthals, who were thought to eat mostly meat, "ate their greens". A tooth analysis revealed that the Europeans hominins consumed mostly cooked and roasted plants. By contrast, the researchers found very few lipids or proteins from meat[439,440].

Likewise, for a long time there was the strong held presumption, in part based on "reports" from 1920, that Arctic populations like the Tikiġaġmiut of the north Alaskan coast or tribes around Canada had based their diet on nearly 100% on the protein and fat of marine mammals and fish. However, investigators identified more than 1000 edible plant species and found evidence that the Arctic inhabitants that gathered, preserved and consumed the majority of them made up a considerable degree in their diet despite the short growing season[441,442].

Looking at the data from rehydrated human coprolites, ancient human feces dropped around 10,000 years ago, revealed that our ancestors appear to have taken in at least 130 grams of high-residue, coarse fiber from plants every day[434,443].

A *Scientific American* blog post by Prof. Ungar puts the debate about the true human diet back to a more scientific point of view: "From the standpoint of paleoecology, the so-called Paleo diet is a myth"[433].

PROTEIN CONCERNS: DO WE GET ENOUGH FROM PLANTS?

The aftermath of World War II and the rise of the food industry in Western countries after the 1960s coincided with a sudden onset of "protein anxiety" among the population.

As a result, the daily protein consumption from animal products per capita has been rising steadily since the 1970s in Western societies such as the United States and western Europe; meanwhile massively exceeding the average daily protein requirement[444,445]. According to official estimations by the Food and Agriculture Organization of the United Nations (FAO) and Organization for Economic Cooperation and Development (OECD) analysts, the global demand for animal protein is further growing at a fast pace including emerging countries with increasing urbanization and rising incomes. Overall, protein consumption is predicted to nearly double globally by 2050, disproportionally overtaking population growth and demand[446].

The concerns around a protein deficiency seem to be even more valid with today's growing popularity of veganism. "Many consumers, say they are afraid that without enough protein they will 'crash', similar to the fear of crashing, or 'bonking', among those who are elite athletes", says Melissa Abbott, Vice President of the consumer research company the Hartman Group confirmed that over the past 25 years nearly 60% of Americans are seriously engaged to increase their protein intake[447].

People overestimate the presumed benefits from an increased protein dose in their diet. "You can eat 300 grams of protein a day, but that doesn't mean you'll put on more muscle than someone who takes in 120 grams a day. You're robbing yourself of other macronutrients that the body needs, like whole grains, fats, and fruits and vegetables", dietitian Jim White comments on behalf of the American Academy of Nutrition and Dietetics in a *New York Times* article[448].

As a matter of fact, people are consuming far more protein than their bodies need. Most American adults eat about 100 grams of protein per day, or roughly twice the officially recommended

amount of about 40–52 grams of protein per day given 60–80 kg
of healthy body weight[447]. The World Health Organization (WHO)
and the European Food Safety Authority (EFSA) recommended
a daily protein intake of 0.66–0.83 grams per kg of normal body
weight, a recommendation applicable even for individuals with
higher protein requirements such as pregnant women[449]. In fact,
there is no disease condition related to protein deficiency affecting
the chronically sick Western societies. As it seems, humanity's
modern protein anxiety is largely unfounded.

While dramatically overestimating our general protein requirements,
we underestimate the protein content of whole plant-based foods
as part of a balanced diet. For example, vegans get on average
60 to 80 grams of protein throughout the day from plant-based
foods including legumes, nuts, broccoli or whole grains. A Swiss
population study on the nutrient intake with different diet patterns,
including a vegan diet, even found that people eating a 100%
plant-based diet had more than 100 grams of protein on average
throughout their day[450]. The researchers concluded "all diets (but not
necessarily all individuals) had adequate protein intake"[450].

According to the official statement from the EFSA, there is
currently no safe upper limit for protein intake, however in contrast
to protein from animal food sources, plant-based protein has not
been documented to generate toxic metabolites in the gut and is
safe to consume.

FEEDING MICROBES FOR DISEASE PREVENTION AND TREATMENT

While nutritional guidelines around the world recommend safe
upper limits for animal-based foods such as meat, there is no safe
upper limit recommendation for any unprocessed plant-based food.

Based on his long-term population studies on chronic disease development, the "fiber man" Dr. Denis Burkitt concluded already in 1974:

> There is no evidence that an increase in cereal fiber in the quantities envisaged, 2 to 6 g of crude fiber a day, could result in any conceivable harm….The betting odds in adopting this procedure would therefore appear to be 'heads I win and tails I don't lose'– not bad odds![224].

In support of Dr. Burkitt's fiber hypothesis, the current advice given by the official Dietary Guidelines for Americans (DGA) for meeting sufficient fiber and nutrient intake in one's daily diet is simple: Increase your daily intake of nutrient-dense foods as much as possible. "Make every bite count"[237].

There is a huge discrepancy between what is officially recommended by guidelines and what most people put on their plates though.

While the average American is concerned about protein intake and exceeds by nearly double, merely 5% of the population achieve the recommended daily intake of fiber-rich plant foods.

Since the 1970s, fiber-poor animal products and processed foods consistently replaced the nutrient-dense, unprocessed plants in people's daily diet increasing the risk for a microbiome starvation and its toxic metabolic shift in noncommunicable chronic diseases.

The daily dietary choices have the most profound impact on the structure and behavior of the human microbiome. While animal foods can contribute to a toxic microbiome metabolism as already discussed earlier, unprocessed plant foods have the power to preserve and even rebuild a healthy microbiome.

Early analyses from 2014 of the relative abundance of taxonomical microbial groups as well as their metabolism have demonstrated that short-term animal-based diets have a greater impact on microbiome community structure and its metabolic changes compared to plant-based diets[198,201]. In comparison to animal protein and fat, the carbohydrate-based plant foods are readily catabolized by the microbiota as it is equipped with up to 60,000 glycoside hydrolase enzymes to turn the fiber into health-promoting short chain fatty acids, suggesting that the human gut microbiome by default favors plant-based diets over animal-based foods[121].

A plant-based diet has consistently been shown to maintain a healthy microbiome and even reverse detrimental metabolic activities in the gut. Consequently, long-term vegetarians and vegans have higher gut microbial diversity and gene richness, their microbes have a better capacity to produce health-promoting short chain fatty acids, their stool pH is lower and their guts are shown to harbor less pathobionts than omnivores[451-456]. The microbiota of long-term vegans also lacks the capacity to produce TMAO. Only after a daily L-carnitine supplement challenge for months, comparing to the amount ingested with a daily steak meal, the TMAO levels in their blood steadily increased as a marker for the microbiome's adaptation to the dietary challenge[364,369,457].

Conversely, a healthy metabolic transformation can be measured in omnivores challenged with a whole food vegan diet. After one month, their fecal enzyme activity changed resulted in a decreased production of the toxic metabolites phenol and p-cresol in their gut[297]. Of note, after resuming their conventional diets, the fecal enzyme activity returned to its previous state within 2 weeks.

It does not even need to be strictly vegan though to benefit from ample plant food intake in a daily diet. Several studies have associated diets high in dietary fiber including significant amounts of fruits, vegetables and whole grains with an increased microbial richness, at either the taxonomic or metabolic level and improved human health, without going strictly vegan.

For example, higher fruit and vegetable consumption in overweight individuals with a low gut microbial gene richness at baseline, which was also characterized by the ability to produce a range of toxic metabolites, increased the subjects' microbiome gene richness and improved their clinical condition[94,458]. Likewise, whole grain cereals, oatmeal and a general increased dietary fiber intake was shown to improve the gut microbial metabolism and a low-grade systemic inflammation even in healthy individuals[459-461,204].

Strategically improving the gut microbiome metabolism by consuming more fiber-rich plant foods can be a potent tool for soothing symptoms of a chronic disease or even reverse it.

For example, patients with inflammatory bowel disease who switch to a plant-based diet regimen produce higher amounts of anti-inflammatory butyrate in their guts and more than 80% of them lived free from an inflammatory relapse for 5 years[273,462-465]. This dietary intervention may be more powerful than the standard of care medication for inflammatory bowel disease, as the usual relapse rate for people is around 40% after 1 year on drugs.

That is why some researchers suggest:

> Although medication is needed in the active phase
> of inflammatory bowel disease, diet is generally more
> important than medication to maintain remission in the

quiescent phase. If a suitable diet is established as part
of a changing lifestyle, medication ultimately may not be
needed to maintain remission[464].

Likewise, a number of studies on type 2 diabetes patients report
improved digestion, increased microbial SCFA production,
diminished toxic compounds such as indole and hydrogen sulfides
and improved clinical parameters after the subjects' dietary fiber
intake was increased[466,467]. The patients had lower hemoglobin
A1c levels, a clinical marker for metabolic syndrome and less
inflammation markers in their blood[467,205].

Increased fiber consumption from whole, plant-based foods may
also affect the gut microbiome indirectly by improving the digestive
process and significantly increasing the intestinal transit time. The
transit time, which is the time it takes for someone from eating food
to subsequent bowel movement, as well as the feces consistency,
have a significant impact on the microbiome. Enriching the diet
of constipated individuals with dietary fiber from whole grains
and fruits, increased their fecal weight and increased their bowel
movements to an intestinal transit time of 24 h to 48 h, which is
currently considered as a normal, healthy digestive pattern[210,468].
Accordingly, when the digestive transit takes longer, lower levels
of healthy microbial metabolites such as SCFAs are produced,
and more toxic metabolites accumulate in the gut lumen[469-472].
Intriguingly, the differing amounts of therapeutic versus toxic
microbial metabolites may be one of the reasons explaining why
women with less frequent bowel movements were shown to have a
higher risk of breast cancer, and possibly colorectal cancer[473,474].

One of the most comprehensive population studies worldwide, the
Global Burden of Disease Study, investigated the major risk factors
for disease and early death across 195 countries.

An unhealthy diet turned out to be the largest global burden of disease and currently poses a greater risk to morbidity and mortality than does unsafe sex, alcohol, drug and tobacco use combined[14,475].

The researchers also evaluated the consumption of the most influential dietary factors and quantified the impact of poor diets on death and disease from noncommunicable diseases in the countries, specifically cancers, cardiovascular diseases and diabetes.

The top spots on the list of the riskiest dietary habits rank all the plant-based foods we forget to eat, followed by animal and processed foods we instead eat too much.

Despite some regional differences, almost no country managed to eat the optimal amount of foods recommended by official guidelines and researchers. For instance, the leading dietary risk factors include a massive underconsumption of fruits, whole grains, nuts and seeds, vegetables, legumes, fiber and omega-3 fatty acids which however is "compensated" for by an excess of salt, red meat, processed meat, sugar-sweetened beverages and trans fatty acids[15].

According to the latest data of the study in 2017, 11 million deaths and 255 million disability-adjusted life years (DALYs) worldwide could have been prevented if we had eaten more nutritious plant-based foods and less fiber-deprived animal and processed foods. A high intake in sodium, too few whole grains and fruits together accounted for more than half of all diet-related deaths globally[15].

The study also elaborated the minimum intake of foods it takes to get all the nutritional benefits for effective protection against noncommunicable diseases. Accordingly, the researchers recommend to eat at least 250 g of fruit, 360 g of fresh vegetables,

60 g of legumes, 125 g of whole grains, 21 g of nuts and seeds per day while limiting the intake of processed meat to 2 g[15].

A high nutritional quality means effective cancer and disease prevention. In one study comparing the nutritional quality of vegan, vegetarian and omnivore diet, researchers awarded the highest index values for diet quality to a vegan diet while the omnivore diet received the lowest[476]. Consequently, plant-based eaters, vegans and vegetarians, are reported to have the lowest total cancer risk, −15% and −8%, respectively, compared with omnivores, as well as the lowest risks of early death from any disease[477,427].

When it comes to diet quality, besides the nutritional density the preservation of the molecular nutrient structure in unprocessed foods seems to be an important yet still underrated factor.

For example, studies show that minimally processed, whole, intact grains, legumes or nuts compared to their processed equivalent (milled or as an ingredient in bread or pastries) were a more readily used substrate for the gut bacteria as their consumption resulted in increased production of SCFAs in people's guts[478,479,221].

This may even translate into a longevity benefit. Whole, intact grains but not refined grains are associated with a lower all-cause mortality by 17% despite comparable fiber[480].

Conversely, the fiber structures contained in processed foods or fiber supplements are usually destroyed, denatured and seem to become a less optimal substrate for the gut microbiota.

Preclinical studies found increased systemic inflammation and harmful secondary bile acid concentration in experimental animals from a processed fiber supplement but not from whole grain chow[481,482]. The researchers even observed that animals were at higher risk to develop liver cancer. Chronic intake of fiber supplements may not only stimulate inflammatory processes and liver cancer but may also "make your gut grow" as an editorial in the journal *Nutrients* titled upon a publication showing an increased cell proliferation in the gut of rats after processed pectin intake[483,484].

The molecular food and fiber structure may be even more crucial for our microbes and our health than the total fiber content. Intact fiber structures in plant foods serve as a scaffold for all the cancer- and disease-preventing antioxidants such as plant polyphenols since they are bound to fiber. Hence, intact fiber can be considered a vehicle carrying essential nutrients into the body which will subsequently be released in the gut with the help from the microbiota's enzymes[485-487]. In contrast, a fiber supplement or processed food is molecularly remodeled and hence lacks the naturally attached phytonutrients. The intake of processed fiber deprives the microbiota of its most vital energy source and may even pose a risk on human long-term health.

While studies provide strong evidence for the disease-protecting and longevity benefits of natural, dietary fiber from unprocessed plant food intake; the evidence for the same benefits from processed and synthetic fiber in humans is inconsistent[488]. Natural plant foods contain an optimal amount of countless disease-protecting nutrients; however, the exact composition and structure of the nutritional components that exert the health benefits are largely undefined, which implies a synergistic effect that is hard to mimic by artificial supplements.

For that reason, the authors of the Global Burden of Disease Study conclude, "This evidence largely endorses a case for moving from nutrient-based to food-based guidelines"[489]. Indeed, the American

Dietetic Association and the National Academy of Sciences Dietary Reference Intake (DRI) committee propose that rather than taking fiber supplements the public should consume adequate amounts of dietary fiber from a variety of plant foods[490-492,237].

Simply by doubling the dietary fiber intake from 15 g to at least 30 g per day, populations could reduce the risk of colorectal cancer by 40%[217]. For example, the colorectal cancer risk drops significantly by merely including a small portion of legumes three times a week and one portion of brown rice at least once a week[493].

A diet centered around plant foods not only has the potential to reduce the risk of cancer, but it may even slow down the progression of cancer.

For example, the prostate-specific antigen (PSA) level in the blood of men diagnosed with prostate cancer randomized to a plant-based diet dropped, whereas the level increased in the control group who continued to eat a standard American diet including animal products[494]. Also, the plant-based men's serum dripped on prostate cancer cell lines *in vitro* inhibited the cancer cell growth by 70%.

The bottom line from all the evidence may be simple, yet powerful: "Eat food. Not too much. Mostly plants", as #1 *New York Times* bestselling author Prof. Michael Pollan puts it[495].

We die if we do not eat plants, but we don't, if we do not eat animals.

The EAT Lancet commission even points out that red meat (processed and unprocessed) "is not essential" and is linearly associated with total mortality, which "without a threshold, suggests that optimal intake would be low"[8]. The experts concluded that an "intake of 0 g/day to about 28 g/day of red meat is desirable, with a midpoint of 14 g/day for the reference diet"[8].

Researchers from the Harvard T.H. Chan School of Public Health and Purdue University found that diets containing high-quality plant protein sources such as legumes, soy and nuts are associated with a markedly improved cardiometabolic profile, compared to diets including red meat[386]. Importantly, simply replacing two servings of animal protein a day with a plant-based protein source significantly lowered the concentration of TMAO in people's serum after 2 months.[387]

In fact, the simple intervention of swapping protein sources may prevent almost 40% of early deaths[496]. In one study, 30 g of mixed nuts per day as part of a Mediterranean diet reduced cardiovascular disease incidents by 28%[497]. The EAT Lancet commission advocates for a daily intake of around 50 g of nuts in place of red meat to protect the heart and save the environment[8].

In a comprehensive literature review on the power of plant-based foods in cancer prevention, the authors Dr. Madigan and Dr. Karhu sum up the case of plant foods against animal foods for an effective chronic disease prevention and a lower risk of dying early: "the safest ratio of animal to plant proteins may be 0:1"[498].

Lessons can be learned from the longest living populations on our planet, the so called "Blue Zone" communities located in different geographical regions across the globe. The "Blue Zone" populations and their secrets to longevity was investigated in 2004 by journalist Dan Buettner, *National Geographic* and longevity researchers. They found that people reach age 100 at rates 10 times greater than the average person in the United States[499]. Five "Blue Zones" were identified: the Adventist community in Loma

Linda in California; the Nicoya Peninsula in Costa Rica; Sardinia in Italy; Ikaria in Greece; and Okinawa in Japan. Apart from various healthy lifestyle factors such as a socially engaging lifestyle, one of the major unifying elements accounting for their longevity was identified to be diet.

The Blue Zone populations consume about 95% of their daily kilojoules from plant foods, with legumes being consumed twice a week in place of meat in Ikaria and almost daily in other Blue Zone areas.

Chapter

8

"Fixing" the Microbiome

Can We Restore a Healthy Microbiome by Other Means Than Diet?

DOI: 10.1201/9781003212447-8

Our microbiome is a lively, ever-changing organ and highly complex ecosystem playing by its own rules. Just like the outer natural environment we live in; our inner microbial environment can be very resilient towards disturbing factors and has the ability to adjust its equilibrium state accordingly. Resilience, however, does not mean immunity towards external forces.

The steady selective pressure from our daily diet constantly "nudges" the microbiome so it adjusts its composition and metabolic behavior in consequence.

While a long-term fiber deprivation leads to microbe starvation and increases the susceptibility for diseases, a balanced plant-based diet lets the microbiota thrive and produce anti-inflammatory, health-promoting metabolites. By our everyday food choices, we have the power to deliberately shape and continuously "educate" our microbiome, ideally in favor of our health.

Abrupt, harsh interventions on the contrary, e.g., antibiotic treatment, cause a sudden and uncontrolled disruption of the microbial ecosystem. Antibiotic treatment results in an acute decimation of the microbial diversity which may take years to recover if at all, with unpredictable long-term consequences for human health as discussed earlier.

Thus, the critical yet justified question arises: Considering the complexity of the microbial ecosystem, does a "quick fix" work to reconstitute a healthy gut microbiome sustainably and thereby prevent or improve chronic disease?

PROBIOTICS – HYPE OR HOPE?

An intuitive approach to repair a dysbiotic gut microbiome, restore missing microbes and increase microbial diversity has gained tremendous enthusiasm among the scientific community and the public for years: probiotics.

In the beginning of the twentieth century, the Russian zoologist Ilya Metchnikoff was the first to claim that ingested probiotic bacteria might be helpful to human health. Based on observations on healthy-aged Bulgarian peasants who regularly consumed fermented milk products he drew the conclusion that aging was due to putrefaction in the colon and that its adverse effects may be counteracted by changing the flora to that of a more saccharolytic metabolism by means of probiotic bacteria ingestion[500].

The demand is gigantic, and hopes are high. According to a business analysis, the global probiotics market is valued at USD 61.1 billion in 2021 and is projected to reach USD 91.1 billion by 2026, at a compound annual growth rate (CAGR) of 8.3% during the forecast period[501].

In 2000, expert panels of the FAO and the WHO addressed the growing interest in probiotics and suggested a uniform definition for the term probiotics: "Live microorganisms, that when administered in adequate amounts, confer a health benefit on the host"[502].

"Could a Bacteria-Stuffed Pill cure Autoimmune Diseases?" is the headline of an article in *Nature* reflecting the huge enthusiasm surrounding probiotics:

> Those might include giving babies well-defined
> compositions of microbes, so that a child's immune
> system develops with optimal tolerance to self without
> sacrificing their ability to fight infection … If we can
> come up with defined compositions of microbes in a
> gummy bear – now we're talking![175]

Is the Hype Justified?

Despite close to 3000 clinical trials in the scientific database on probiotics (by the end of 2021) trying to show an improvement in various chronic disease conditions, the results concerning their efficacy are inconsistent, largely conflicting and thus dampen the enthusiasm instead calling for a thorough benefit for cost evaluation.

Generally, despite the excitement about an inverse relationship between probiotic intake and an improvement for some patients as concluded in some studies, a large number of studies were unable to show a clear benefit noting that the "totality of evidence has been deemed insufficient to support a therapeutic claim" and "the effectiveness of the administration of probiotics at a clinical level remains elusive"[503–505].

Even researchers who clearly advocate for probiotic administration acknowledge the caveat that the "quality of evidence is still limited and requires further investigation"[506].

So far, probiotic treatments have shown limited effect in some human conditions including antibiotic-associated diarrhea, ulcerative colitis, inflammatory bowel syndrome, improved health in neonates or prevention of atopic dermatitis in children.

For instance, randomized controlled trials suggest that probiotics may reduce the time to full enteral feed in preterm and low-birth weight neonates by 1.5 days[507]. One prospective study claimed that perinatal administration of a probiotic strain was associated with reduced risk of atopic dermatitis, however, the probiotic was additionally supplemented with retinol, calcium and zinc, implicating a confounding effect[508]. In contrast, a randomized, placebo-controlled trial of probiotics for primary prevention of dermatitis in children could not show a benefit, revealing that atopic dermatitis was diagnosed in 28% of the subjects in the probiotics group versus 27.3% in the placebo group[509].

With this in mind, another approach was suggested to support the development of the microbiome and thereby regulating a child's immune system: dirt. A biodiversity intervention trial published in *Science* demonstrated that children exposed to a wider spectrum of environmental microbes on their skin and mouth mucosa in nature-oriented daycare centers improved their immune system's ability to reduce inflammatory pathways in their body compared to children in a standard urban daycare environment[83]. The researchers propose: "For this reason, one can anticipate that providing children with a chance for daily contact with diverse vegetation and dirt in safe urban green spaces such as playgrounds and daycare center or schoolyards might improve child health by activating the regulatory pathways of the immune system". They conclude that this accessible and safe measure might have better therapeutic effects than oral treatment with probiotic bacteria.

Also, although probiotics are commonly used for the prevention of antibiotic-associated diarrhea (AAD), the optimum regimen remains controversial. Of 51 publications only 6 probiotic therapies showed some efficacy compared with a placebo[510]. Only one strain, *Lactobacillus rhamnosus* GG, was ranked as more effective than others.

One article analyzing 24 randomized controlled trials on the efficacy of probiotics in inducing a remission or preventing relapse in inflammatory bowel disease (IBD) even found a worse outcome for people on probiotics in comparison to standard of care medication in terms of efficacy and side effects. There were 31.6%

randomized patients on probiotics that failed to achieve remission, compared with 25.4% receiving 5-ASAs (Mesalazine)[511]. Adverse events occurred in 15.8% patients assigned to probiotics compared with 11.9% of those allocated to 5-ASAs. Even when comparing to a placebo, the probiotic demonstrated no or even a worse result for treating the inflammation and a higher percentage of patients had side effects. Merely one specific probiotic mixture, VSL#3, appeared to show a slight benefit in inducing a remission for 26% of ulcerative colitis patients[511]. No benefit was shown for Crohn's disease though, leaving the vast majority of patients with unfulfilled hopes and unmet needs.

As a result, a quite unusual yet effective approach may deem more suitable and to recommend to IBD patients from doctors: eating more fiber-rich, plant-based foods. A plant-based diet was effective in preventing a relapse in around 80% of patients for up to 5 years[273,464]. For example, legume and potato consumption are inversely associated with disease relapse in patients that ate the most, carrying a 79% lower risk of active disease[512].

Similarly, despite a tremendous number of studies, to date only limited evidence exists for a benefit from probiotics for people suffering from irritable bowel syndrome (IBS), a functional gastrointestinal disorder affecting up to 10–20% of the population worldwide.

Generally, researchers detected a huge heterogeneity between studies and a significant publication bias. While some studies were able to show slight improvement in symptoms, for 8–25% of the patients compared to no intervention or a placebo, there were also a large number of studies showing no benefit over placebo at all[513,514]. Which species, strains or the combination thereof might be effective for IBS, however, also remains elusive.

A sustainable and effective approach may be to balance the gut microbial ecosystem naturally by feeding the microbiota appropriately with dietary fiber.

Although a small amount of 7 g of a soluble fiber supplement per day was shown to help IBS patients to increase their fecal *bifidobacteria* and significantly improve their symptom score, a chronic use of fiber supplements may have an adverse effect on gut health by leading to increased cell proliferation[484,515]. In contrast, the daily intake of 2 servings of fiber-rich fruits may effectively alleviate digestive symptoms in patients without side effects. For example, randomized controlled trials concluded that 300 g (2 servings) per day of fresh mangos, or 5 g of mango fiber, significantly reduced symptoms by 60% in chronically constipated adults compared to baseline. Similar results had been observed for kiwi fruits and prunes[516].

Despite the uncertainties there is an overwhelming variety of probiotic supplements in stores leaving customers with a tremendous shopping challenge and unsatisfying low value.

Why Is There So Much Inconsistency in Probiotics Research?

Apart from technical constrains affecting the product quality and efficacy including wide differences concerning the origin and combination of microbial strains, their included amount, duration and conditions of intake, industrial production or storage conditions, there is another neglected inherent feature of the human microbiome that may explain the varying efficacy of commercial probiotic bacteria: the colonization resistance.

The concept of colonization resistance is the ability of commensal microbiota to prevent the overgrowth of pathogenic bacteria[173,517]. This seems to apply to all non-inherent microbes though, no matter if "bad" or "good" bacteria, including probiotics. In fact, the introduction of a non-indigenous microorganism can be considered as a "biological invasion" into the resident microbial community"[517].

The inability of commercial probiotic strains to persist in the human gut might result from their lack of key traits necessary to successfully compete in the gut microbial ecosystem. For instance, most probiotic strains currently used in supplements or foods belong to bacterial taxa that are allochthonous to the human gastrointestinal tract[518]. Dr. Tami Lieberman, a researcher working at the Massachusetts Institute of Technology (MIT), explains: "Within-person evolution may in part explain the stability of the microbiome and why probiotics so infrequently colonize healthy individuals; indigenous, preadapted microbiota may be better suited to the unique combination of selective forces within each individual"[519].

In fact, only after long-term probiotic treatment with one single bacterium known to be a core member of the human gut microbiota already anyways, still only 30% of the study participants adapted the ingested probiotic strain, at least transiently, given a nutritional substrate for the bacterium was present[518].

Several scientists raise complaints and express doubts about the true benefit of probiotics in the treatment of chronic diseases. "While their health benefits seem intuitive and the supportive in vitro evidence is promising, it is doubtful that the benefits of probiotics and prebiotics can outweigh those of a normal 'balanced' diet that our genome evolved together with", as gastroenterologist Dr. Kishore Vipperla argues[236].

The European Food Safety Authority (EFSA) points out to be cautious about healthy claims on microorganisms in probiotics and food due to insufficient evidence for a direct benefit on human physiology[520]. Researchers at a workshop hosted by the esteemed National Academy of Sciences agreed that "Metchnikoff's observations were useful to the dairy food industry, and several fermented dairy products were marketed during the last century, claiming to favor a good microbial balance in the gut"[125]. Although some ingested probiotics may survive within the gastrointestinal tract – albeit in low quantities – "unfortunately, only carefully selected strains, and not those generally contained in yogurt, are able to colonize the gut" doubting not only probiotic supplementation to be effective but also making a case against food products containing live microorganisms[125].

Furthermore, caution is required when evaluating probiotics efficacy: Several meta-analyses from randomized controlled trials detected a significant publication bias in probiotic studies favoring the outcome towards a certain probiotic manufacturer in addition to common label fraught with only around 40% of the products being in accordance with their label claims[513,521,522]. Therefore, some scientists call for "a global comprehensive legislation to control the quality of probiotics whose market is gaining huge momentum"[521].

Researchers also raise awareness about serious health risks and adverse events resulting from probiotics treatment. Although "generally regarded as safe" (GRAS) by safety agencies, probiotics have been reported to be not so safe at all: in studies probiotic treatment caused a delay of the gut microbiota recovery after antibiotic treatment, induced systemic infections in cancer patients, stimulated an excessive immune response in susceptible individuals, led to metabolic acidosis and toxic secondary bile acid production in the gut and was even suspected to transfer antibiotic-resistant genes[523–525].

Worrisome, adverse effects associated with probiotics consumption are widely under-reported in clinical trials further complicating the debate about efficacy[526]. Some researchers demand that "the factors that must be considered in assessing the safety of probiotic products should include infectivity, pathogenicity, an excessive immune stimulation in susceptible individuals, virulence factors comprising toxicity, metabolic activity and the important properties of microbes".

Conclusively, the potential safety risk, financial burden, lack of evidence for a clear benefit and the risk of transfer of antibiotic resistance genes "may be an argument against use of probiotics" according to researchers[525].

However, in the future sound research may prove otherwise.

FECAL MICROBIAL TRANSPLANTATION OR "THE POWER OF POOP"

Instead of administering a pill stuffed with a small mix of randomly combined bacteria to a patient, researchers also came up with the idea to replace the microbiome as a whole akin to transplanting a kidney: a fecal microbial transplantation (FMT).

FMT defines as the administration of fecal matter solution from a healthy donor into the intestinal tract of a recipient[527].

Intriguingly, transplanting the stool of healthy donors to treat gastrointestinal (GI) diseases has ancient origins. A fourth-century Chinese physician gave the "yellow soup" to patients to treat diarrhea[527]. The first successful clinical transplantation was performed in 1958 in the United States in a patient suffering from a severe colon infection. Since then it proved to be mostly effective in around 90% of patients with recurrent *Clostridioides difficile* (*C. difficile*) infections, which is the only condition in which FMT is approved as a therapy by the FDA[528-530].

Does an unappealing fecal "smoothie" applied to the patient via rectal and nasal gastric tube or a "poo pill" soon turn out to be the next-generation probiotic to restore a sick patient's microbiome?

Apart from the astonishing efficacy for *C. difficile* infections, other, albeit small clinical trials, have shown some effect on disease improvement with IBS or ulcerative colitis being the best studied examples. For instance, although 56% of IBS patients with a transplanted stool reported improvement, so did 26% of patients given placebo "poo"[531].

Likewise, a meta-analysis of RCTs analyzed that healthy stool transplants induced clinical remission in 28% of ulcerative colitis patients compared with 9% of patients given a placebo, deeming FMT to be a feasible option for patients to relieve the inflammation[532].

Comparing to the low efficacy of a standard probiotic pill these promising results raise high expectations in sufferers. The promising treatment effect however may vanish soon. After a year more than half of the responders with inflammatory bowel syndrome were back to their symptoms[531].

There are also a few single case reports of autism, Parkinson's or immunotherapy refractory melanoma patients claiming an improvement after an FMT treatment. In the latter case, the cancer patients had an ameliorated response to cancer drug treatment, suggesting that the fresh dose of microbiota in the patients' guts stimulated their immune system to recognize and support fighting the cancer cells[533,534].

According to current evidence, the efficacy of a whole microbiome transplant seems to be based on the two main premises that (a) patients with microbiome dysbiosis have either completely lost their healthy microbiota or (b) that their microbiota is unable to regain its normal functionality, which predominantly apply to most cases of *C. difficile* infection and, albeit to a lesser extent, for ulcerative colitis patients. However, these premises are not necessarily given for other conditions or every patient. Concerning the discussion about the efficacy of FMT, one article in *Science* notes:

> That's akin to having a field decimated by herbicides and fertilizers and expecting it to grow a healthy crop. Cancer patients typically still have their own indigenous, intact microbiome, which is more like an established prairie that an oncologist is trying to overseed with some beneficial species[535].

The human microbiome is an ecosystem. The stability of the ecosystem may be more fragile in patients compared with healthy individuals, yet its own equilibrium state may be strong enough to reject alien microbes and poses an obstacle to the efficacy of fecal transplantation akin to probiotic supplementation. In contrast to a probiotic supplement containing mostly autochthonous bacteria that meet strong resistance from the patient's own microbes, a fecal transplant from a healthy donor contains a broad spectrum of naturally occurring species which however may more easily hold foot. Therefore, a high dose of human-borne species in natural combination may serve at least a severely affected patient, fulfilling the afore-mentioned premise better than a probiotic to reconstitute a healthy microbiome. For example, preclinical studies confirm that FMT but not a selection of single, independent probiotic species induced a rapid microbiome recovery after antibiotic treatment[523,536].

The reports about the superb efficacy of fecal matter transplants have given people with hope that there could be an easy way to treat the disease. Scientists share this desire but warn that clinical research has barely begun. Problems of standardization are accompanied by potentially incalculable long-term risks, many of which are inherent to the transfer of living microorganisms. Researchers point out that

> although FMT also has been shown to be safe and efficacious in immunocompromised patients, it still would be highly desirable to reduce the risk of adverse events in patients with limited eligibility for FMT. Moreover, even the most rigorous and costly donor screening procedures, or defined panels of bacteria, cannot exclude the risk of transferring unknown pathogens or undetectable functional characteristics within the living microorganisms to the recipient, including bacterial or viral risk factors for metabolic diseases, cancer, atopy, or autoimmunity[537].

Previously, the safety of the transplantation procedure was called into question in 2019 when researchers published an article in the *New England Journal of Medicine* describing a case in which a donor stool contaminated with drug-resistant *E. coli* left one man dead and another severely ill[538]. Another study also reported seven cases of patients who received a stool transplant from a donor colonized with Shiga toxin-producing *Escherichia coli* (STEC), luckily nobody died[539].

Improved donor stool screening and procedure conduct are required to prevent further such tragic, adverse events.

Since long-term data are lacking, more research is needed to verify the efficacy of the procedure, elaborate other disease conditions that may benefit and finally grant patients a safe application.

THE "WILD WEST" OF MICROBIOME SCIENCE: DRUGGING THE MICROBIOME AND PERSONALIZED NUTRITION

The idea of a targeted modification of the gut microbiome to treat diseases was not only discussed with tremendous excitement by scientists. Big Pharma and biotech industries also jumped on the train smelling a newborn business opportunity to "drug" the microbiome.

The growing market of therapeutics targeting the microbiome comprises prebiotics, probiotics, postbiotics, "poop pills", biologics to probiotic skin care and vaginal microbiota transplants. An editorial in the journal *Nature* states sarcastically:

> Various companies offer testing of your (or your dog's) gut microbiome and then make personalized diet and lifestyle recommendations or sell supplements. And for the so inclined there are now even smart toilets that send faecal bacterial counts directly to your smartphone. As a microbiologist, one cannot help but think of some of these developments as the Wild West of microbiome science[108].

As discussed earlier, a diet high in animal products and low in plant-based foods is a risk factor for disrupting the microbial ecosystem, which weakens the intestinal barrier and results in the production of toxic metabolites such as TMAO contributing to the development of chronic diseases.

Instead of advising to remove the key drivers causing the toxic processes in the body in the first place, namely meat, dairy and eggs, researchers and companies propose to pharmacologically block the signaling pathways leading to TMAO, in order to prevent the negative outcomes of the low-fiber/animal product-rich diet[285,540]. "This study shows for the first time that one can target a gut microbial pathway to inhibit atherosclerosis. This new approach opens the door to the concept of drugging the microbiome to affect a therapeutic benefit in the host" as the authors in the related publication claim[540].

Intriguingly, research and industry alike recognize and acknowledge the tremendous power of diet as the master modulator of the human gut microbiome. Concerned about the differences between individuals' microbiota though, they argue that targeted therapeutical manipulation of the microbiome likely requires a more "personalized" approach: precision diet.

"Because both host genetics and the gut microbiome can influence host phenotype and treatment outcome, there is an urgent need to develop precision medicine and personalize dietary supplementation based on an individual's microbiome" points Professor Wan of the University of California out[541].

Some public health actors have begun to view large-scale precision nutrition as a novel opportunity to provide the right dietary intervention to the right population at the right time. Current trials attempt to uncover the magnitude of effect the microbiome has on the postprandial metabolic response independent from macronutrients for example. Although they figured out huge interindividual differences in metabolic response to an identical meal, the actual data revealed that the subjects' specific microbiomes hardly seem to have an impact on blood fat content or glycemic response compared to the meal macronutrients itself[542,543]. One of the studies found that at most, 14% of individual differences in postprandial glucose response may be explained by individual differences in microbiota compositions[543]. Nevertheless, the authors confidently claim: "the intestinal microbiome is a co-determinant of the postprandial plasma glucose response". They suggest that "one might be able to recommend a diet higher in animal protein and lower in starchy carbohydrates", while neglecting that rather than starchy vegetables, whole grains or legumes, the high amount of animal products in most diets of industrialized populations are identified to be the major culprit for driving the epidemic of chronic diseases[105].

At the same time the researchers acknowledge that their approach is all but ripe for the clinic "… with overall microbiome features explaining relatively little of the variation of glycemic indexes … (they) might still not be ready for implementation in precision nutrition platforms"[105].

Since the human holobiont is a complex, living system that is subject to constant change, scientists are also worried that "what is 'optimal' may change over time and thus require constant re-evaluation"[105].

Current data on precision diets raise concerns among scientists about the actual rationale and the practicability of advocating for this approach among the general population or confronting the already overtaxed health care systems.

Regarding that the major health burden of modern societies are diet-related chronic diseases accounting for more than two-thirds of years of life lost (YLL) globally, critical voices begin to question the costly, complicated precision diet approach as a valid solution for the global public health epidemic. "Great future or greedy venture: Precision medicine needs philosophy", as the authors of one publication demand[544]. They criticize that "perhaps, medicine in the molecular era might be no more 'precise' than in prior eras. At present, the problem for PM (precision medicine) is more absence of evidence than evidence of absence, indicating the present foundation and framework of PM is not sufficiently sound"[544].

For example, one of the most prominent studies on the effect of precision nutrition published in *Cell* by Dr. Zeevi and colleagues claimed there is high interpersonal variability in postprandial glycaemic responses (PPGRs), that personal and microbiome features enable accurate glucose response prediction that is superior to common practice and that short-term personalized dietary interventions successfully lower post-meal glucose[545]. Soon after, the study was accused to be based on a flawed rationale, as the definition and use of the terms "postprandial glycaemic response" and "variability" was unclear and imprecise[546]. In addition, the majority of the variation in normalized glucose response could actually be accounted for by intra-individual variation but did not explain large inter-individual variation. Most importantly, the authors showed that using their precision nutrition prediction model to reduce postprandial glycaemic response in individuals was no better than general advice to eliminate the foods that are already known to elicit high glucose response. The mean reduction in people's postprandial glycaemic response using the author's precision nutrition model was 46% which is similar to that for general expert-based advice 44%[547]. One editorial in the *European Journal of Clinical Nutrition* sums up: "Zeevi et al., contribute some interesting and novel findings; however, their results do not demonstrate high interpersonal variation in relative glycaemic responses, do not show that their model is superior to current methods of detecting hyperglycaemia, and do not show that personalized nutrition advice is superior to standard dietary advice to manage high post-prandial glucose"[546].

Public health researchers fear that precision nutrition may likely have a limited positive impact on the rising epidemic of obesity or type 2 diabetes at a population level, as it neglects the causes of chronic diseases which are rooted in the obesogenic food environment. They even caution the manifestation of a two-tier health care system, as precision nutrition may be adopted "more likely among more socially privileged members of the population, which would exacerbate socioeconomic inequalities in diet and obesity"[548].

Despite all the exciting new treatment options and business opportunities for industries, the question remains, if our societies can counteract the rising trend of chronically sick people, cope with increasing cancer rates and substantial economic burden on health care systems with complicated, unsafe and costly interventions.

Simplification with maximum results may be key. Especially when it comes to diet proven to be the most powerful tool to prevent chronic diseases and cancer. A healthy diet should be accessible and affordable for every human being rather than an insurmountable obstacle for many. For a diet to be health promoting, satisfying and nurturing it does not need to be overly "precise" at all. The Sonnenburg Lab suggests "It is possible that metabolites produced by commonly occurring or retained pathways, such as SCFAs, may be sufficiently represented in modernized microbial configurations. If so, a general, population-wide, non-personalized approach could have a large positive health impact"[65].

Our growing understanding about how the human microbiome works as a vivid micro-ecosystem and its significance in human health and disease development indicates that a dysbiotic microbiome may not be simply repaired by artificial modifications. In fact, current data do not support a "quick fix" to rebuild a healthy microbiome.

Ultimately, it may be so simple as differentiating between healthy SCFA-inducing foods (plants), and potentially toxic metabolite triggering foods (animal protein and saturated fat) and choosing a healthy balance.

Human evolution and the large body of current scientific evidence confirm that the most nurturing, safe, effective, economic, accessible, sustainable and natural way to modify the microbiome to act in a health-promoting manner is simpler than we may have thought: by following a plant-based diet centering around whole grains, legumes, nuts, vegetables and fruits while keeping fiber-poor animal based and processed foods at bay. What made most of us sick in the first place can ultimately be our cure: Our dietary choices.

Eating a hot dog could cost you 36 minutes of a healthy life, while eating a serving of nuts instead could help you gain 26 minutes[10].

You choose.

GLOSSARY

16S rRNA 16S ribosomal RNA (or 16S rRNA) is the RNA component of the 30S subunit of a prokaryotic ribosome. The 16S rRNA gene is used for phylogenetic studies as it is highly conserved between different species of bacteria and archaea[549]

bacterial phylum while the exact definition of a bacterial phylum is debated, a popular definition is that a bacterial phylum is a monophyletic lineage of bacteria whose 16S rRNA genes share a pairwise sequence identity of ~75% or less with those of the members of other bacterial phyla[550]

bacterial taxa bacteria classified by a hierarchy of ranks (species, genus, family, order, class, phylum)[551]

Bacteroidetes gram-negative bacterial phylum of the gut microbiota

bile acids primary bile acids are an essential component of the process of solubilization and digestion of dietary lipids. Bile acids are hydroxylated steroids, synthesized in the liver from cholesterol.

butyrate a four-carbon short chain fatty acid (SCFA), is produced through microbial fermentation of dietary fibers in the lower intestinal tract. Endogenous butyrate production, delivery, and absorption by colonocytes have been well documented. Butyrate exerts its functions by acting as a histone deacetylase (HDAC) inhibitor or signaling through several G protein–coupled receptors (GPCRs). Recently, butyrate has received particular attention for its beneficial effects on intestinal homeostasis and energy metabolism. With anti-inflammatory properties, butyrate enhances intestinal barrier function and mucosal immunity

choline choline is a quaternary ammonium salt and a
precursor of phospholipids, such as sphingomyelin and
phosphatidylcholine (lecithin), which are used to build
the cell membrane during the cell proliferation process.
Choline is a nutrient in the vitamin B complex. Contained
in high amounts in egg yolks

colonocytes epithelial cell of the colon

dysbiosis change in gut microbiota metabolism from human
health promoting to potentially disease promoting

enteroptype stable clusters of gut microbiota. enterotypes are
mostly driven by species composition, but without providing
molecular functional traits[552]

fecal microbial transplant FMT defines as the administration
of fecal matter solution from a healthy donor into the
intestinal tract of a recipient

fiber nondigestible carbohydrates for humans that make up the
plant cell wall and can be fermented by the gut microbiota
and converted into mostly short chain fatty acids[553]

Firmicutes gram-positive bacterial phylum of the gut microbiota

germ-free mice germ-free mice are bred in isolators which fully block
exposure to microorganisms, with the intent of keeping them
free of detectable bacteria, viruses, and eukaryotic microbes.
Germ-free mice allow for study of the complete absence
of microbes or for the generation of gnotobiotic animals
exclusively colonized by known microbes[554]

heme iron dietary iron comes in two forms: heme iron and
non-heme iron. Iron, is a primary component of hemoglobin,
a protein in red blood cells that carries oxygen to all
parts of the body and part of myoglobin, a protein that
transports and stores oxygen in the muscles. heme iron is a
predominant component of animal tissues

holobiont "host-symbiont" or composite organism composed of
human and microbial cells[555]

intestinal barrier the intestinal barrier is a semipermeable
    structure that allows the uptake of essential nutrients and
    immune sensing, while being restrictive against pathogenic
    molecules and bacteria. Both structural and molecular
    components act together to fulfil this complex, but essential
    function of the gastrointestinal tract

intestinal crypt the epithelium of the small intestine is organized
    into large numbers of self-renewing crypt-villus units. The
    base of each villus is surrounded by multiple epithelial
    invaginations, termed crypts of Lieberkühn. Crypts are
    home to a population of vigorously proliferating epithelial
    cells, which fuel the active self-renewal of the epithelium[556]

intestinal epithelium the layer of intestinal epithelial cells (IECs)
    provides a physical and biochemical barrier that segregates
    host tissue and commensal bacteria to maintain intestinal
    homeostasis

L-carnitine L-carnitine is a key component of the so-called
    carnitine shuttle, a multienzyme transport system that is
    required to transfer activated long chain fatty acids
    (acyl-CoAs) into the mitochondrial matrix, where they are
    degraded via β-oxidation. L-carnitin is contained in meat[557]

lecithin phosphatidylcholine, an ubiquitous dietary lipid

metabolism the chemical changes in living cells by which energy is
    provided for vital processes and activities and new material
    is assimilated

metabolomics metabolomics identifies and determines the set of
    metabolites (or specific metabolites) in biological samples
    (tissues, cells, fluids or organisms) under normal conditions
    in comparison with altered states promoted by disease,
    drug treatment, dietary intervention, or environmental
    modulation[558]

metagenomics metagenomics is the study of all of the genes from
    many different organisms in a population. In terms of the

human gut microbiome, this process not only provides
detailed information about the bacteria strains present but
also indicates the enhancing capabilities of those different
strains, based on their genetics, to keep the gut in good
working order

metaproteomics metaproteomics can be defined as large-scale
characterization of the entire protein complement of
environmental microbiota at a given point in time

microbe microorganism, a diverse group of generally minute
simple life-forms that include bacteria, archaea, algae,
fungi, protozoa and viruses.

microbiome (1) a community of microorganisms (such as bacteria,
fungi, and viruses) that inhabit a particular environment
and especially the collection of microorganisms living in or
on the human body

(2) the collective genomes of microorganisms inhabiting
a particular environment and especially the human body

microbiota a community of microorganisms (such as bacteria,
fungi and viruses) that inhabit a particular environment
and especially the collection of microorganisms living in or
on the human body

N-nitroso compounds NOC compounds include both nitrosamines
and nitrosamides and can participate in DNA alkylation
which can lead to tumor formation. Animal experiments
have demonstrated N-nitroso compounds to be the
most potent and broadly acting carcinogens known.
Several hundred N-nitroso compounds have been tested
as carcinogens and over 80% have been found to be
carcinogenic in at least 40 animal models[559,560]

niche (a) habitat supplying the factors necessary for the existence
of an organism or species

(b) the ecological role of an organism in a community
especially in regard to food consumption

noncommunicable diseases noncommunicable diseases (NCDs), also known as chronic diseases, tend to be of long duration and are the result of a combination of genetic, physiological, environmental and behavioral factors. The main types of NCD are cardiovascular diseases (such as heart attacks and stroke), cancers, chronic respiratory diseases (such as chronic obstructive pulmonary disease and asthma) and diabetes[561]

non-heme iron dietary iron comes in two forms: heme iron and non-heme iron. Iron is a primary component of hemoglobin, a protein in red blood cells that carries oxygen to all parts of the body and part of myoglobin, a protein that transports and stores oxygen in the muscles. non-heme iron is a predominant component of plants such as whole grains, nuts and seeds, legumes and leafy greens. the natural bioavailability is lower than heme-iron but is multiplied when consumed with small amounts of vitamin C-rich foods

operational taxonomical units an OTU can be defined as a collection of 16S rRNA sequences that have a certain percentage of sequence divergence[562]

probiotics a product or preparation that contains live microorganisms, that when administered in adequate amounts, confer a health benefit on the host[502]

processed foods according to the US Department of Agriculture, processed foods are any raw agricultural commodities that have been washed, cleaned, milled, cut, chopped, heated, pasteurized, blanched, cooked, canned, frozen, dried, dehydrated, mixed or packaged – anything done to them that alters their natural state. This may include adding preservatives, flavors, nutrients and other food additives, or substances approved for use in food products, such as salt, sugars and fats

putrefaction anaerobic catabolism of proteins by bacteria and
          fungi with the formation of foul-smelling incompletely
          oxidized products

secondary bile acids primary bile acids that enter the colon
          can be metabolized by the bacterial flora and converted
          into secondary bile acids, deoxycholic acid (DCA) and
          lithocholic acid (LCA), respectively

short chain fatty acids (SCFAs) short chain fatty acids including
          butyrate, propionate or acetate are the main metabolic
          products of anaerobic bacterial fermentation in the
          intestine. In addition to their important role as fuel
          for intestinal epithelial cells, SCFAs modulate different
          processes in the gastrointestinal (GI) tract and in other
          tissues such as adipose and immune tissues

trimethylamine N-oxide trimethylamine N-oxide (TMAO) is a small
          colorless amine oxide generated from choline, betaine, and
          carnitine by gut microbial metabolism. It accumulates in the
          tissue of marine animals in high concentrations and protects
          against the protein-destabilizing effects of urea. Plasma level
          of TMAO is determined by a number of factors including
          diet, gut microbial flora and liver flavin monooxygenase
          activity. In humans, a positive correlation between elevated
          plasma levels of TMAO and an increased risk for major
          adverse cardiovascular events and death is reported[563]

xenobiotics a xenobiotic is defined as a chemical that is not used
          by the reference organism as a nutrient chemical, is not
          essential to the reference organism for maintenance of
          normal physiologic/biochemical function and homeostasis,
          and does not constitute a part of the conventional array
          of chemicals synthesized from nutrient chemicals by the
          reference organism in normal intermediary metabolism[564]

LIST OF ABBREVIATIONS

BA	primary bile acid
DCA	deoxycholic acid (secondary bile acid)
DDT	dichloro-diphenyl-trichloroethane (insecticide)
DGA	Dietary Guidelines for Americans
DOC	deoxycholate (secondary bile acid)
EDC	endocrine-disrupting chemicals
ENS	enteric nervous system
FAO	Food and Agriculture Organization of the United Nations
GWAS	genome-wide association study
H₂S	hydrogen sulfide
HCA	heterocyclic amines
HMP	Human Microbiome Project
IBD	inflammatory bowel disease
IBS	inflammatory bowel syndrome
iHMP	Integrative Human Microbiome Project
LCA	lithocholic acid (secondary bile acid)
LPS	lipopolysaccharide
NCD	noncommunicable diseases
NOC	N-nitroso compounds
OECD	Organisation for Economic Co-operation and Development
OTU	operational taxonomic unit
PCBs	polychlorinated biphenyls
POP	persistent organic pollutants
SAA	sulfur amino acid
SAD	Standard American Diet
SCFA	short chain fatty acid

SDG   Sustainable Development Goals
TMA   trimethylamine
TMAO   trimethylamine N-oxide
WCRF   World Cancer Research Fund
WHO   World Health Organization

LITERATURE

1. NIH. 2020 NCI Budget Fact Book – Research Funding – National Cancer Institute. https://www.cancer.gov/about-nci/budget/fact-book/data/research-funding. Accessed April 3, 2022.

2. NCD countdown 2030: Pathways to achieving Sustainable Development Goal Target 3.4. *Lancet*. 2020;396(10255):918–934. doi:10.1016/S0140-6736(20)31761-X

3. Noncommunicable diseases. https://www.who.int/news-room/fact-sheets/detail/noncommunicable-diseases. Accessed November 24, 2021.

4. The Cancer Atlas. The Burden of Cancer | The Cancer Atlas. https://canceratlas.cancer.org/the-burden/the-burden-of-cancer/. Accessed March 24, 2022.

5. WHO. Noncommunicable diseases prematurely take 16 million lives annually, WHO urges more action. *WHO*. 2015.

6. WHO. Mortality and global health estimates. https://www.who.int/data/gho/data/themes/mortality-and-global-health-estimates. Accessed April 1, 2022.

7. Sender R, Fuchs S, Milo R. Revised estimates for the number of human and bacteria cells in the body. *bioRxiv*. 2016.

8. Willett W, Rockström J, Loken B, et al. Food in The Anthropocene: The EAT–Lancet commission on healthy diets from sustainable food systems. *Lancet*. 2019;393(10170):447–492. doi:10.1016/S0140-6736(18)31788-4

9. Visconti A, Le Roy CI, Rosa F, et al. Interplay between the human gut microbiome and host metabolism. *Nat Commun.* 2019;10(1):1–10. doi:10.1038/s41467-019-12476-z

10. Stylianou KS, Fulgoni VL, Jolliet O. Small targeted dietary changes can yield substantial gains for human health and the environment. *Nat Food 2021 28.* 2021;2(8):616–627. doi:10.1038/s43016-021-00343-4

11. Disability-adjusted life years (DALYs). https://www.who.int/data/gho/indicator-metadata-registry/imr-details/158. Accessed November 26, 2020.

12. Wengler A, Rommel A, Plaß D, et al. Years of life lost to death. *Dtsch Arztebl Int.* 2021;118(9):137–144. doi:10.3238/arztebl.m2021.0148

13. The disease burden of noncommunicable diseases. PAHO/WHO | Pan American Health Organization. https://www.paho.org/en/noncommunicable-diseases-and-mental-health/enlace-data-portal-noncommunicable-diseases-mental-0. Accessed February 10, 2022.

14. Lozano R, Naghavi M, Foreman K, et al. global and regional mortality from 235 causes of death for 20 age groups in 1990 and 2010: A systematic analysis for the Global Burden of Disease Study 2010. *Lancet.* 2012;380(9859):2095–2128. doi:10.1016/S0140-6736(12)61728-0

15. Afshin A, Sur PJ, Fay KA, et al. Health effects of dietary risks in 195 countries, 1990–2017: A systematic analysis for the Global Burden of Disease Study 2017. *Lancet.* 2019;393(10184):1958–1972. doi:10.1016/S0140-6736(19)30041-8

16. Lang T, Heasman M. Food Wars: The Global Battle for Mouths, Minds and Markets. Google Books. https://books.google.se/books?id=1eO9CgAAQBAJ&pg=PA30&lpg=PA30&d

q=post+war+productionism&source=bl&ots=CVWfkAWqST
&sig=ACfU3U1rYYtsIVGuRvxzDmT0KsboINa5Vg&hl=en&sa=
X&ved=2ahUKEwi53fOplvj1AhUMlIsKHb9fCI0Q6AF6BAgRE
AM#v=onepage&q=post war productionism&f=false. Accessed
February 11, 2022.

17. Kearney J. Food consumption trends and drivers. *Philos Trans
R Soc B Biol Sci.* 2010;365(1554):2793–2807. doi:10.1098/
rstb.2010.0149

18. FAOSTAT. https://www.fao.org/faostat/en/#data. Accessed
November 24, 2021.

19. Breakdown and projection of worldwide consumption based
on type of food... | Download Scientific Diagram. https://
www.researchgate.net/figure/Breakdown-and-projection-of-
worldwide-consumption-based-on-type-of-food-Source-FAO_
fig2_335960589. Accessed September 26, 2021.

20. Popkin BM. The nutrition transition and its health
implications in lower-income countries. *Public Health Nutr.*
1998;1(1):5–21. doi:10.1079/phn19980004

21. Whitnall T, Pitts N. *Agricultural Commodities.* March 2019.

22. OECD-FAO Agricultural Outlook 2021–2030. July 2021.
doi:10.1787/19991142

23. An Overview of Meat Consumption in the United States.
https://farmdocdaily.illinois.edu/2021/05/an-overview-
of-meat-consumption-in-the-united-states.html. Accessed
November 24, 2021.

24. Yeh MC, Glick-Bauer M, Wechsler S. Fruit and vegetable
consumption in the United States: Patterns, barriers and
federal nutrition assistance programs. *Fruits, Veg Herbs
Bioact Foods Heal Promot.* January 2016:411–422. doi:10.1016/
B978-0-12-802972-5.00019-6

25. *The State of Food Security and Nutrition in the World 2020.* FAO, IFAD, UNICEF, WFP and WHO; 2020. doi:10.4060/ca9692en

26. Allen L, De Benoist B, Dary O, Hurrell R. Guidelines on food fortification with micronutrients. Allen L, de Benoist B, Dary O, Hurrell R, eds. *Guidel Food Fortif with Micronutr.* 2006:3–20. doi:10.3/JQUERY-UI.JS

27. World hunger is still not going down after three years and obesity is still growing. UN Report. https://www.who.int/news/item/15-07-2019-world-hunger-is-still-not-going-down-after-three-years-and-obesity-is-still-growing-un-report. Accessed June 24, 2021.

28. Worldwide trends in diabetes since 1980: A pooled analysis of 751 population-based studies with 4.4 million participants. *Lancet (London, England).* 2016;387(10027):1513–1530. doi:10.1016/S0140-6736(16)00618-8

29. Obesity and Overweight. https://www.who.int/news-room/fact-sheets/detail/obesity-and-overweight. Accessed June 21, 2021.

30. Health and Economic Costs of Chronic Diseases. CDC. https://www.cdc.gov/chronicdisease/about/costs/index.htm. Accessed November 24, 2021.

31. Fact Sheets. https://www.who.int/news-room/fact-sheets. Accessed November 24, 2021.

32. Safiri S, Sepanlou SG, Ikuta KS, et al. The global, regional, and national burden of colorectal cancer and its attributable risk factors in 195 countries and territories, 1990–2017: A systematic analysis for the Global Burden of Disease Study 2017. *Lancet Gastroenterol Hepatol.* 2019;4(12):913–933. doi:10.1016/S2468-1253(19)30345-0

33. American Cancer Society. Colorectal Cancer Statistics. https://www.cancer.org/cancer/colon-rectal-cancer/about/key-statistics.html. Accessed February 11, 2022.

34. O'Keefe SJD, Li J V., Lahti L, et al. Fat, fibre and cancer risk in African Americans and rural Africans. *Nat Commun.* 2015;6:6342. doi:10.1038/ncomms7342

35. Stemmermann GN, Nomura AMY, Chyou P -H, Kato I, Tetsuo T. Cancer incidence in Hawaiian Japanese: Migrants from Okinawa compared with those from other prefectures. *Jpn J Cancer Res.* 1991;82(12):1366–1370. doi:10.1111/J.1349-7006.1991.TB01807.X

36. Kaplan GG. The global burden of IBD: From 2015 to 2025. *Nat Rev Gastroenterol Hepatol.* 2015;12(12):720–727. doi:10.1038/NRGASTRO.2015.150

37. Ananthakrishnan AN. Environmental triggers for inflammatory bowel disease. *Curr Gastroenterol Rep.* 2013;15(1):302. doi:10.1007/s11894-012-0302-4

38. Jantchou P, Morois S, Clavel-Chapelon F, Boutron-Ruault M-C, Carbonnel F. Animal protein intake and risk of inflammatory bowel disease: The E3N prospective study. *Am J Gastroenterol.* 2010;105(10):2195–2201. doi:10.1038/ajg.2010.192

39. WHO. *NCD Report WHO.*

40. Anand P, Kunnumakara AB, Sundaram C, et al. Cancer is a preventable disease that requires major lifestyle changes. *Pharm Res.* 2008;25(9):2097–2116. doi:10.1007/s11095-008-9661-9

41. WCRF/AICR Score. *Diet, Nutrition, Physical Activity and Colorectal Cancer.*

42. World Cancer Research Fund UK. Higher number of bowel and breast cancer cases can be prevented through lifestyle changes. https://www.wcrf-uk.org/uk/latest/press-releases/higher-number-bowel-and-breast-cancer-cases-can-be-prevented-through. Accessed June 20, 2021.

43. WCRF. WHO Nutrition goals and SDG 2030 unlikely to be met. https://www.wcrf.org/wp-content/uploads/2021/02/EB148-Agenda-Item-6-Nutrition-Version-for-websites.pdf. Published 2021. Accessed June 9, 2021.

44. Margulis L. Symbiosis as a Source of Evolutionary Innovation. MIT Press. https://mitpress.mit.edu/books/symbiosis-source-evolutionary-innovation. Accessed February 11, 2022.

45. Cummings JH, Macfarlane GT. Role of intestinal bacteria in nutrient metabolism. *JPEN J Parenter Enteral Nutr.* 1997;21(6):357–365. doi:10.1177/0148607197021006357

46. Moeller AH, Caro-Quintero A, Mjungu D, et al. Cospeciation of gut microbiota with hominids. *Science.* 2016;353(6297):380–382. doi:10.1126/SCIENCE.AAF3951

47. Tito RY, Knights D, Metcalf J, Obregon-Tito AJ, Cleeland L. Insights from characterizing extinct human gut microbiomes. *PLoS One.* 2012;7(12):51146. doi:10.1371/journal.pone.0051146

48. Tito RY, Macmil S, Wiley G, et al. Phylotyping and functional analysis of two ancient human microbiomes. *PLoS One.* 2008;3(11):e3703. doi:10.1371/JOURNAL.PONE.0003703

49. Cano RJ, Rivera-Perez J, Toranzos GA, et al. Paleomicrobiology: Revealing fecal microbiomes of ancient indigenous cultures. *PLoS One.* 2014;9(9). doi:10.1371/journal.pone.0106833

50. Adler CJ, Dobney K, Weyrich LS, et al. Sequencing ancient calcified dental plaque shows changes in oral microbiota with dietary shifts of the neolithic and industrial revolutions. *Nat Genet.* 2013;45(4):450–455. doi:10.1038/NG.2536

51. Wibowo MC, Yang Z, Borry M, et al. Reconstruction of ancient microbial genomes from the human gut. *Nat 2021 5947862.* 2021;594(7862):234–239. doi:10.1038/s41586-021-03532-0

52. Harvard Medical School. The Guts of Our Ancestors. https://hms.harvard.edu/news/guts-our-ancestors?utm_source=Silverpop&utm_medium=email&utm_term=field_news_item_1&utm_content=HMNews05172021. Accessed December 2, 2021.

53. De Filippo C, Cavalieri D, Di Paola M, et al. Impact of diet in shaping gut microbiota revealed by a comparative study in children from Europe and rural Africa. *Proc Natl Acad Sci USA.* 2010;107(33):14691–14696. doi:10.1073/pnas.1005963107

54. Mancabelli L, Milani C, Lugli GA, et al. Meta-analysis of the human gut microbiome from urbanized and pre-agricultural populations. *Environ Microbiol.* 2017;19(4):1379–1390. doi:10.1111/1462-2920.13692

55. Obregon-Tito AJ, Tito RY, Metcalf J, et al. Subsistence strategies in traditional societies distinguish gut microbiomes. *Nat Commun.* 2015;6. doi:10.1038/ncomms7505

56. García-Vega ÁS, Corrales-Agudelo V, Reyes A, Escobar JS. Diet quality, food groups and nutrients associated with the gut microbiota in a nonwestern population. *Nutrients.* 2020;12(10):1–21. doi:10.3390/nu12102938

57. García-Ortiz H, Barajas-Olmos F, Contreras-Cubas C, et al. The genomic landscape of Mexican indigenous populations brings insights into the peopling of the Americas. *Nat Commun*. 2021;12(1):5942. doi:10.1038/S41467-021-26188-W

58. Gomez A, Petrzelkova KJ, Burns MB, et al. Gut microbiome of coexisting BaAka pygmies and Bantux reflects gradients of traditional subsistence patterns. *Cell Rep*. 2016;14(9):2142–2153. doi:10.1016/j.celrep.2016.02.013

59. Schnorr SL, Candela M, Rampelli S, et al. Gut microbiome of the Hadza hunter-gatherers. *Nat Commun*. 2014;5. doi:10.1038/ncomms4654

60. Amato KR, Metcalf JL, Song SJ, et al. Using the gut microbiota as a novel tool for examining colobine primate GI health. *Glob Ecol Conserv*. 2016;7:225–237. doi:10.1016/j.gecco.2016.06.004

61. Clayton JB, Vangay P, Huang H, et al. Captivity humanizes the primate microbiome. *Proc Natl Acad Sci U S A*. 2016;113(37):10376–10381. doi:10.1073/PNAS.1521835113

62. Ryan MJ, Schloter M, Berg G, et al. Development of microbiome biobanks – Challenges and opportunities. *Trends Microbiol*. 2021;29(2):89–92. doi:10.1016/j.tim.2020.06.009

63. Dominguez Bello MG, Knight R, Gilbert JA, Blaser MJ. Preserving microbial diversity. *Science*. 2018;362(6410):33–34. doi:10.1126/SCIENCE.AAU8816

64. Microbiota Vault. https://www.microbiotavault.org/. Accessed December 2, 2021.

65. Sonnenburg ED, Sonnenburg JL. The ancestral and industrialized gut microbiota and implications for human health. *Nat Rev Microbiol*. 2019;17(6):383–390. doi:10.1038/s41579-019-0191-8

66. Moeller AH, Li Y, Ngole EM, et al. Rapid changes in the gut microbiome during human evolution. *Proc Natl Acad Sci U S A.* 2014;111(46):16431–16435. doi:10.1073/PNAS.1419136111/-/DCSUPPLEMENTAL

67. *The Convergence of Infectious Diseases and Noncommunicable Diseases.* National Academies Press; 2019. doi:10.17226/25535

68. Sonnenburg JL, Sonnenburg ED. Vulnerability of the industrialized microbiota. *Science.* 2019;366(6464). doi:10.1126/science.aaw9255

69. Blaser MJ. Missing microbes: How the overuse of antibiotics is fueling our modern plagues; 273. https://books.google.com/books/about/Missing_Microbes.html?hl=de&id=iB5OAwAAQBAJ. Accessed December 9, 2021.

70. Russell SL, Gold MJ, Hartmann M, et al. Early life antibiotic-driven changes in microbiota enhance susceptibility to allergic asthma. *EMBO Rep.* 2012;13(5):440. doi:10.1038/EMBOR.2012.32

71. Marra F, Marra CA, Richardson K, et al. Antibiotic use in children is associated with increased risk of asthma. *Pediatrics.* 2009;123(3):1003–1010. doi:10.1542/PEDS.2008-1146

72. Palleja A, Mikkelsen KH, Forslund SK, et al. Recovery of gut microbiota of healthy adults following antibiotic exposure. *Nat Microbiol.* 2018;3(11):1255–1265. doi:10.1038/s41564-018-0257-0

73. Cully M. Antibiotics alter the gut microbiome and host health. *Nat Res.* June 2019. https://www.nature.com/articles/d42859-019-00019-x. Accessed December 9, 2021.

74. Dethlefsen L, Relman DA. Incomplete recovery and individualized responses of the human distal gut microbiota to repeated antibiotic perturbation. *Proc Natl Acad Sci.* 2011;108(Supplement_1):4554–4561. doi:10.1073/pnas.1000087107

75. Vatanen T, Kostic AD, D'Hennezel E, et al. Variation in microbiome LPS immunogenicity contributes to autoimmunity in humans. *Cell.* 2016;165(4):842–853. doi:10.1016/j.cell.2016.04.007

76. Muller LMAJ, Gorter KJ, Hak E, et al. Increased risk of common infections in patients with type 1 and type 2 diabetes mellitus. *Clin Infect Dis.* 2005;41(3):281–288. doi:10.1086/431587

77. Kaspersen KA, Pedersen OB, Petersen MS, et al. Obesity and risk of infection: Results from the Danish blood donor study. *Epidemiology.* 2015;26(4):580–589. doi:10.1097/EDE.0000000000000301

78. Thaiss CA, Levy M, Grosheva I, et al. Hyperglycemia drives intestinal barrier dysfunction and risk for enteric infection. *Science.* 2018;359(6382):1376–1383. doi:10.1126/science.aar3318

79. Jakobsson HE, Abrahamsson TR, Jenmalm MC, et al. Decreased gut microbiota diversity, delayed bacteroidetes colonisation and reduced Th1 responses in infants delivered by caesarean section. *Gut.* 2014;63(4):559–566. doi:10.1136/GUTJNL-2012-303249

80. Bertelsen RJ, Lise Brantsaeter A, Haugen M, et al. The impact of caesarian section on the relationship between inhalent allergen exposure and allergen-specific IgE at age 2 years. *J Allergy Clin Immunol.* 2013;131(2):AB129. doi:10.1016/J.JACI.2012.12.1129

81. Darabi B, Rahmati S, Hafeziahmadi MR, Badfar G, Azami M. The association between caesarean section and childhood asthma: An updated systematic review and meta-analysis. *Allergy, Asthma Clin Immunol.* 2019;15(1). doi:10.1186/s13223-019-0367-9

82. Laubereau B, Filipiak-Pittroff B, Von Berg A, et al. Caesarean section and gastrointestinal symptoms, atopic dermatitis, and sensitisation during the first year of life. *Arch Dis Child.* 2004;89(11):993. doi:10.1136/ADC.2003.043265

83. Roslund MI, Puhakka R, Grönroos M, et al. Environmental studies biodiversity intervention enhances immune regulation and health-associated commensal microbiota among daycare children. *Sci Adv.* 2020;6(42). doi:10.1126/SCIADV.ABA2578/SUPPL_FILE/ABA2578_SM.PDF

84. Wilkins LJ, Monga M, Miller AW. Defining dysbiosis for a cluster of chronic diseases. *Sci Rep.* 2019;9(1). doi:10.1038/s41598-019-49452-y

85. Weir TL, Manter DK, Sheflin AM, Barnett BA, Heuberger AL, Ryan EP. Stool microbiome and metabolome differences between colorectal cancer patients and healthy adults. *PLoS One.* 2013;8(8). doi:10.1371/JOURNAL.PONE.0070803

86. Thomas AM, Manghi P, Asnicar F, et al. Metagenomic analysis of colorectal cancer datasets identifies cross-cohort microbial diagnostic signatures and a link with choline degradation. *Nat Med.* 2019;25(4):667–678. doi:10.1038/s41591-019-0405-7

87. Davies J. In a map for human life, count the microbes, too. *Science.* 2001;291(5512):2316b–2316. doi:10.1126/science.291.5512.2316b

88. Relman DA, Falkow S. The meaning and impact of the human genome sequence for microbiology. *Trends Microbiol.* 2001;9(5):206–208. doi:10.1016/S0966-842X(01)02041-8

89. Peterson J, Garges S, Giovanni M, et al. The NIH human microbiome project. *Genome Res.* 2009;19(12):2317–2323. doi:10.1101/gr.096651.109

90. Falony G, Joossens M, Vieira-Silva S, et al. Population-level analysis of gut microbiome variation. *Science.* 2016;352(6285):560–564. doi:10.1126/science.aad3503

91. Jackson MA, Verdi S, Maxan ME, et al. Gut microbiota associations with common diseases and prescription medications in a population-based cohort. *Nat Commun 2018 91.* 2018;9(1):1–8. doi:10.1038/s41467-018-05184-7

92. Riquelme E, Zhang Y, Zhang L, et al. Tumor microbiome diversity and composition influence pancreatic cancer outcomes. *Cell.* 2019;178(4):795–806.e12. doi:10.1016/J.CELL.2019.07.008

93. Ciernikova S, Mego M, Novisedlakova M, Cholujova D, Stevurkova V. The emerging role of microbiota and microbiome in pancreatic ductal adenocarcinoma. *Biomedicines.* 2020;8(12):1–21. doi:10.3390/biomedicines8120565

94. Le Chatelier E, Nielsen T, Qin J, et al. Richness of human gut microbiome correlates with metabolic markers. *Nature.* 2013;500(7464):541–546. doi:10.1038/nature12506

95. Cani PD, Van Hul M. Microbial signatures in metabolic tissues: A novel paradigm for obesity and diabetes? *Nat Metab.* 2020;2(3):211–212. doi:10.1038/s42255-020-0182-0

96. Zhou W, Sailani MR, Contrepois K, et al. Longitudinal multi-omics of host–microbe dynamics in prediabetes. *Nature.* 2019;569(7758):663–671. doi:10.1038/s41586-019-1236-x

97. Pedersen HK, Gudmundsdottir V, Nielsen HB, et al. Human gut microbes impact host serum metabolome and insulin sensitivity. *Nature.* 2016;535(7612):376–381. doi:10.1038/nature18646

98. Mottawea W, Chiang CK, Mühlbauer M, et al. Altered intestinal microbiota–host mitochondria crosstalk in new onset Crohn's disease. *Nat Commun.* 2016;7(1):1–14. doi:10.1038/ncomms13419

99. Joossens, M. Huys, G. Cnockaert, M. et al. Dysbiosis of the faecal microbiota in patients with Crohn's disease and their unaffected relatives. *Gut.* 2011;60(5):631–637. doi:10.1136/GUT.2010.223263

100. Cekanaviciute E, Yoo BB, Runia TF, et al. Gut bacteria from multiple sclerosis patients modulate human T cells and exacerbate symptoms in mouse models. *Proc Natl Acad Sci U S A.* 2017;114(40):10713–10718. doi:10.1073/PNAS.1711235114/-/DCSUPPLEMENTAL

101. Wirbel J, Pyl PT, Kartal E, et al. Meta-analysis of fecal metagenomes reveals global microbial signatures that are specific for colorectal cancer. *Nat Med.* 2019;25(4):679–689. doi:10.1038/s41591-019-0406-6

102. Schluter J, Peled JU, Taylor BP, et al. The gut microbiota is associated with immune cell dynamics in humans. *Nature.* 2020;588(7837):303–307. doi:10.1038/s41586-020-2971-8

103. Peled JU, Gomes ALC, Devlin SM, et al. Microbiota as predictor of mortality in allogeneic hematopoietic-cell transplantation. *N Engl J Med.* 2020;382(9):822–834. doi:10.1056/NEJMOA1900623

104. Manor O, Dai CL, Kornilov SA, et al. Health and disease markers correlate with gut microbiome composition across thousands of people. *Nat Commun.* 2020;11(1). doi:10.1038/s41467-020-18871-1

105. Bartek N, Blumberg J, Blander G, Jorge M. Gut Microbiota-informed precision nutrition in the generally healthy individual: Are we there yet? *Curr Dev Nutr.* 2021;5(9):nzab107–nzab107. doi:10.1093/CDN/NZAB107

106. Hanage WP. Microbiology: Microbiome science needs a healthy dose of scepticism. *Nature.* 2014;512(7514):247–248. doi:10.1038/512247a

107. Walter J, Armet AM, Finlay BB, Shanahan F. Establishing or exaggerating causality for the gut microbiome: Lessons from human microbiota-associated rodents. *Cell.* 2020;180(2):221–232. doi:10.1016/J.CELL.2019.12.025

108. Hype or hope? *Nat Rev Microbiol.* 2019;17(12):717–717. doi:10.1038/s41579-019-0283-5

109. Jones S. Trends in microbiome research. *Nat Biotechnol.* 2013;31(4):277. doi:10.1038/NBT.2546

110. Kroes I, Lepp PW, Relman DA. Bacterial diversity within the human subgingival crevice. *Proc Natl Acad Sci.* 1999;96(25):14547–14552. doi:10.1073/PNAS.96.25.14547

111. Markets and Markets. Human Microbiome Market—Global Forecast to 2028. https://www.marketsandmarkets.com/Market-Reports/human-microbiome-market-37621904.html. Accessed December 15, 2021.

112. Jonathan Eisen auf Twitter. The Tree of Life: Overselling the Microbiome Award: The Microbiome Diet Book. http://t.co/1YHglrBg6q"/Twitter. https://twitter.com/phylogenomics/status/504297373296640000. Accessed February 14, 2022.

113. Marchesi JR, Ravel J. The vocabulary of microbiome research: A proposal. *Microbiome.* 2015;3(1). doi:10.1186/s40168-015-0094-5

114. Vujkovic-Cvijin I, Sklar J, Jiang L, Natarajan L, Knight R, Belkaid Y. Host variables confound gut microbiota studies of human disease. *Nature.* 2020;587(7834):448–454. doi:10.1038/s41586-020-2881-9

115. Lloréns-Rico V, Raes J. Tracking humans and microbes. *Nature.* 2019;569(7758):632–633. doi:10.1038/d41586-019-01591-y

116. Thomas, L. The Lives of a Cell: Notes of a Biology Watcher. Google Books. https://books.google.de/books/about/The_Lives_of_a_Cell.html?id=pFRtrv-_fpAC&redir_esc=y. Accessed June 28, 2021.

117. Bosch TCG, McFall-Ngai MJ. Metaorganisms as the new frontier. *Zoology (Jena).* 2011;114(4):185–190. doi:10.1016/J.ZOOL.2011.04.001

118. Whipps JM. Mycoparasitism and plant disease control. *Fungi Biol Control Syst.* 1988:161–187.

119. Microbiome—Definition & Meaning. Merriam-Webster. https://www.merriam-webster.com/dictionary/microbiome. Accessed February 15, 2022.

120. Berg G, Rybakova D, Fischer D, et al. Microbiome definition re-visited: Old concepts and new challenges. *Microbiome.* 2020;8(1):103. doi:10.1186/s40168-020-00875-0

121. Kaoutari A El, Armougom F, Gordon JI, Raoult D, Henrissat B. The abundance and variety of carbohydrate-active enzymes in the human gut microbiota. *Nat Rev Microbiol.* 2013;11(7):497–504. doi:10.1038/nrmicro3050

122. Gillois K, Lévêque M, Théodorou V, Robert H, Mercier-Bonin M. Mucus: An underestimated gut target for environmental pollutants and food additives. *Microorganisms.* 2018;6(2):53. doi:10.3390/microorganisms6020053

123. Madhusoodanan J. Inner workings: Microbiota munch on medications, causing big effects on drug activity. *Proc Natl Acad Sci.* 2020;117(17):9135–9137. doi:10.1073/PNAS.2003785117

124. Bäckhed F, Ley RE, Sonnenburg JL, Peterson DA, Gordon JI. Host-bacterial mutualism in the human intestine. *Science.* 2005;307(5717):1915–1920. doi:10.1126/science.1104816

125. Ending the War Metaphor. *End War Metaphor.* June 2006. doi:10.17226/11669

126. Bocci V. The neglected organ: Bacterial flora has a crucial immunostimulatory role. *Perspect Biol Med.* 1992;35(2):251–260. doi:10.1353/pbm.1992.0004

127. Wang Z, Ying Z, Bosy-Westphal A, et al. Specific metabolic rates of major organs and tissues across adulthood: Evaluation by mechanistic model of resting energy expenditure. *Am J Clin Nutr.* 2010;92(6):1369–1377. doi:10.3945/AJCN.2010.29885

128. Wang Z, Ying Z, Bosy-Westphal A, et al. Evaluation of specific metabolic rates of major organs and tissues: Comparison between nonobese and obese women. *Obesity.* 2012;20(1): 95–100. doi:10.1038/OBY.2011.256

129. Scott A. Homing in on the molecules from microbes. *Nature.* 2020;577(7792):S9. doi:10.1038/d41586-020-00195-1

130. De Vadder F, Kovatcheva-Datchary P, Goncalves D, et al. Microbiota-generated metabolites promote metabolic benefits via gut-brain neural circuits. *Cell.* 2014;156(1-2). doi:10.1016/j.cell.2013.12.016

131. Koh A, Vadder F De, Kovatcheva-Datchary P, Bäckhed F. From dietary fiber to host physiology: Short-chain fatty acids as key bacterial metabolites. *Cell*. 2016;165(6):1332–1345. doi:10.1016/J.CELL.2016.05.041

132. Frost G, Sleeth ML, Sahuri-Arisoylu M, et al. The short-chain fatty acid acetate reduces appetite via a central homeostatic mechanism. *Nat Commun*. 2014;5:3611. doi:10.1038/ncomms4611

133. Absorption of short-chain fatty acids by the colon. PubMed. https://pubmed.ncbi.nlm.nih.gov/6768637/. Accessed December 21, 2021.

134. Le Poul E, Loison C, Struyf S, et al. Functional characterization of human receptors for short chain fatty acids and their role in polymorphonuclear cell activation. *J Biol Chem*. 2003;278(28):25481–25489. doi:10.1074/jbc.M301403200

135. Cummings JH, Pomare EW, Branch HWJ, Naylor CPE, MacFarlane GT. Short chain fatty acids in human large intestine, portal, hepatic and venous blood. *Gut*. 1987;28(10):1221–1227. doi:10.1136/GUT.28.10.1221

136. Sun Q, Jia Q, Song L, Duan L. Alterations in fecal short-chain fatty acids in patients with irritable bowel syndrome: A systematic review and meta-analysis. *Med (United States)*. 2019;98(7). doi:10.1097/MD.0000000000014513

137. Leccioli V, Oliveri M, Romeo M, Berretta M, Rossi P. A New Proposal for the Pathogenic Mechanism of Non-Coeliac/Non-Allergic Gluten/Wheat Sensitivity: Piecing Together the Puzzle of Recent Scientific Evidence. 2017. doi:10.3390/nu9111203

138. Magnusson MK, Isaksson S, Öhman L. The anti-inflammatory immune regulation induced by butyrate is impaired in inflamed intestinal mucosa from patients with ulcerative colitis. *Inflammation*. 2020;43(2):507–517. doi:10.1007/s10753-019-01133-8

139. Deleu S, Machiels K, Raes J, Verbeke K, Vermeire S. Short chain fatty acids and its producing organisms: An overlooked therapy for IBD? *EBioMedicine*. 2021;66. doi:10.1016/J.EBIOM.2021.103293

140. Chambers ES, Viardot A, Psichas A, et al. Effects of targeted delivery of propionate to the human colon on appetite regulation, body weight maintenance and adiposity in overweight adults. *Gut*. 2015;64(11):1744–1754. doi:10.1136/gutjnl-2014-307913

141. Mörkl S, Lackner S, Meinitzer A, et al. Gut microbiota, dietary intakes and intestinal permeability reflected by serum zonulin in women. *Eur J Nutr*. 2018;57(8):2985–2997. doi:10.1007/S00394-018-1784-0

142. Li Q, Cao L, Tian Y, et al. Butyrate suppresses the proliferation of colorectal cancer cells via targeting pyruvate kinase M2 and metabolic reprogramming. *Mol Cell Proteomics*. 2018;17(8):1531–1545. doi:10.1074/MCP.RA118.000752/ATTACHMENT/8D67EA70-5BB5-4404-99C3-9F8CE1EBAB6E/MMC1.ZIP

143. Szentirmai É, Millican NS, Massie AR, Kapás L. Butyrate, a metabolite of intestinal bacteria, enhances sleep. *Sci Rep*. 2019;9(1):1–9. doi:10.1038/s41598-019-43502-1

144. Kim SW, Hooker JM, Otto N, et al. Whole-body pharmacokinetics of HDAC inhibitor drugs, butyric acid, valproic acid and 4-phenylbutyric acid measured with carbon-11 labeled analogs by PET. *Nucl Med Biol*. 2013;40(7):912–918. doi:10.1016/J.NUCMEDBIO.2013.06.007

145. Frost G, Sleeth ML, Sahuri-Arisoylu M, et al. The short-chain fatty acid acetate reduces appetite via a central homeostatic mechanism. *Nat Commun*. 2014;5. doi:10.1038/NCOMMS4611

146. Li Z, Yi CX, Katiraei S, et al. Butyrate reduces appetite and activates brown adipose tissue via the gut-brain neural circuit. *Gut.* 2018;67(7):1269–1279. doi:10.1136/GUTJNL-2017-314050

147. Madsen HB, Ahmed SH. Drug versus sweet reward: Greater attraction to and preference for sweet versus drug cues. *Addict Biol.* 2015;20(3):433–444. doi:10.1111/adb.12134

148. Sekikawa A, Higashiyama A, Lopresti BJ, et al. Associations of equol-producing status with white matter lesion and amyloid-β deposition in cognitively normal elderly Japanese. *Alzheimer's Dement Transl Res Clin Interv.* 2020;6(1). doi:10.1002/TRC2.12089

149. Mora C, Danovaro R, Loreau M. Alternative hypotheses to explain why biodiversity-ecosystem functioning relationships are concave-up in some natural ecosystems but concave-down in manipulative experiments. *Sci Reports 2014 41.* 2014;4(1):1–9. doi:10.1038/srep05427

150. Gutenberg TPJD. Devotions Upon Emergent Occasions by John Donne. https://www.gutenberg.org/files/23772/23772-h/23772-h.htm. Published 1624. Accessed March 7, 2022.

151. Boets E, Gomand S V., Deroover L, et al. Systemic availability and metabolism of colonic-derived short-chain fatty acids in healthy subjects: A stable isotope study. *J Physiol.* 2017;595(2):541–555. doi:10.1113/JP272613

152. den Besten G, Lange K, Havinga R, et al. Gut-derived short-chain fatty acids are vividly assimilated into host carbohydrates and lipids. *Am J Physiol – Gastrointest Liver Physiol.* 2013;305(12). doi:10.1152/ajpgi.00265.2013

153. Sachs JL, Hollowell AC. The origins of cooperative bacterial communities. *MBio.* 2012;3(3). doi:10.1128/mBio.00099-12

154. Eckburg PB, Bik EM, Bernstein CN, et al. Microbiology: Diversity of the human intestinal microbial flora. *Science.* 2005;308(5728):1635–1638. doi:10.1126/science.1110591

155. Woo M. Mapping the microbiome location helps elucidate its role. *Proc Natl Acad Sci U S A.* 2018;115(48):12078–12080. doi:10.1073/pnas.1816174115

156. Welch JLM, Rossetti BJ, Rieken CW, Dewhirst FE, Borisy GG. Biogeography of a human oral microbiome at the micron scale. *Proc Natl Acad Sci U S A.* 2016;113(6):E791–E800. doi:10.1073/PNAS.1522149113

157. Welch JLM, Hasegawa Y, McNulty NP, Gordon JI, Borisy GG. Spatial organization of a model 15-member human gut microbiota established in gnotobiotic mice. *Proc Natl Acad Sci U S A.* 2017;114(43):E9105–E9114. doi:10.1073/pnas.1711596114

158. Tanoue T, Morita S, Plichta DR, et al. A defined commensal consortium elicits CD8 T cells and anti-cancer immunity. *Nat 2019 5657741.* 2019;565(7741):600–605. doi:10.1038/s41586-019-0878-z

159. Turnbaugh PJ, Quince C, Faith JJ, et al. Organismal, genetic, and transcriptional variation in the deeply sequenced gut microbiomes of identical twins. *Proc Natl Acad Sci U S A.* 2010;107(16):7503–7508. doi:10.1073/pnas.1002355107

160. Zhernakova A, Kurilshikov A, Bonder MJ, et al. Population-based metagenomics analysis reveals markers for gut microbiome composition and diversity. *Science.* 2016;352(6285). doi:10.1126/science.aad3369

161. De Martel C, Ferlay J, Franceschi S, et al. Global burden of cancers attributable to infections in 2008: A review and synthetic analysis. *Lancet Oncol.* 2012;13(6):607–615. doi:10.1016/S1470-2045(12)70137-7

162. Biological agents. Volume 100 B. A review of human carcinogens. PubMed. https://pubmed.ncbi.nlm.nih. gov/23189750/. Accessed March 10, 2021.

163. Institute of Medicine. Microbial threats to health: Emergence, detection, and response. *Microb Threat to Heal.* March 2003. doi:10.17226/10636

164. Lee YC, Chiang TH, Chou CK, et al. Association between *Helicobacter pylori* eradication and gastric cancer incidence: A systematic review and meta-analysis. *Gastroenterology.* 2016;150(5):1113–1124.e5. doi:10.1053/J.GASTRO.2016.01.028

165. Unidentified curved bacilli on gastric epithelium in active chronic gastritis. PubMed. https://pubmed.ncbi.nlm.nih. gov/6134060/. Accessed January 4, 2022.

166. Linz B, Balloux F, Moodley Y, et al. An African origin for the intimate association between humans and *Helicobacter pylori*. *Nature.* 2007;445(7130):915–918. doi:10.1038/NATURE05562

167. Barzegari A, Saeedi N, Saei AA. Shrinkage of the human core microbiome and a proposal for launching microbiome biobanks. *Future Microbiol.* 2014;9(5):639–656. doi:10.2217/ fmb.14.22

168. Azuma T, Suto H, Ito Y, et al. Eradication of *Helicobacter pylori* infection induces an increase in body mass index. *Aliment Pharmacol Ther Suppl.* 2002;16(2):240–244. doi:10.1046/j.1365-2036.16.s2.31.x

169. Francois F, Roper J, Joseph N, et al. The effect of *H. pylori* eradication on meal-associated changes in plasma ghrelin and leptin. *BMC Gastroenterol.* 2011;11. doi:10.1186/1471-230X-11-37

170. Erőss B, Farkas N, Vincze Á, et al. *Helicobacter pylori* infection reduces the risk of Barrett's esophagus: A meta-analysis and systematic review. *Helicobacter.* 2018;23(4). doi:10.1111/hel.12504

171. Kasai C, Sugimoto K, Moritani I, et al. Changes in plasma ghrelin and leptin levels in patients with peptic ulcer and gastritis following eradication of *Helicobacter pylori* infection. *BMC Gastroenterol.* 2016;16(1). doi:10.1186/s12876-016-0532-2

172. Zhou X, Wu J, Zhang G. Association between *Helicobacter pylori* and asthma: A meta-analysis. *Eur J Gastroenterol Hepatol.* 2013;25(4):460–468. doi:10.1097/MEG.0b013e32835c280a

173. Clemente JC, Ursell LK, Parfrey LW, Knight R. The impact of the gut microbiota on human health: An integrative view. *Cell.* 2012;148(6):1258–1270. doi:10.1016/j.cell.2012.01.035

174. Koopen AM, de Clercq NC, Warmbrunn M V., et al. Plasma metabolites related to peripheral and hepatic insulin sensitivity are not directly linked to gut microbiota composition. *Nutrients.* 2020;12(8):1–12. doi:10.3390/NU12082308

175. Bender E. Could a bacteria-stuffed pill cure autoimmune diseases? *Nature.* 2020;577(7792):S12–S13. doi:10.1038/D41586-020-00197-Z

176. Armour CR, Nayfach S, Pollard KS, Sharpton TJ. A metagenomic meta-analysis reveals functional signatures of health and disease in the human gut microbiome. Segata N, ed. *mSystems.* 2019;4(4). doi:10.1128/mSystems.00332-18

177. Rojas OL, Pröbstel AK, Porfilio EA, et al. Recirculating intestinal IgA-producing cells regulate neuroinflammation via IL-10. *Cell.* 2019;176(3):610–624.e18. doi:10.1016/J.CELL.2018.11.035

178. Pianta A, Arvikar SL, Strle K, et al. Two rheumatoid arthritis-specific autoantigens correlate microbial immunity with autoimmune responses in joints. *J Clin Invest.* 2017;127(8):2946–2956. doi:10.1172/JCI93450

179. Franzosa EA, Huang K, Meadow JF, et al. Identifying personal microbiomes using metagenomic codes. *Proc Natl Acad Sci U S A.* 2015;112(22):E2930–E2938. doi:10.1073/pnas.1423854112

180. Callaway E. Microbiomes raise privacy concerns. *Nature.* 2015;521(7551):136. doi:10.1038/521136A

181. Bouslimani A, Melnik A V, Xu Z, et al. Lifestyle chemistries from phones for individual profiling. *Proc Natl Acad Sci.* 2016;113(48):E7645–E7654. doi:10.1073/PNAS.1610019113

182. Teuer und sinnlos: DGVS rät von Stuhltests zur Analyse des Darm-Mikrobioms ab - DGVS - Deutsche Gesellschaft für Gastroenterologie, Verdauungs und Stoffwechselkrankheiten. https://www.dgvs.de/pressemitteilungen/teuer-und-sinnlos-dgvs-raet-von-stuhltests-zur-analyse-des-darm-mikrobioms-ab/. Accessed May 12, 2020.

183. Adams D, Wroe N, Cerf C, Dawkins R. The salmon of doubt: Hitchhiking the galaxy one last time. Google Books. https://books.google.com/books/about/The_Salmon_of_Doubt.html?hl=de&id=2nJZrOrvyNIC. Accessed March 7, 2022.

184. Bell T. Next-generation experiments linking community structure and ecosystem functioning. *Environ Microbiol Rep.* 2019;11(1):20–22. doi:10.1111/1758-2229.12711

185. Long S, Yang Y, Shen C, et al. Metaproteomics characterizes human gut microbiome function in colorectal cancer. *Npj Biofilms Microbiomes.* 2020;6(1). doi:10.1038/S41522-020-0123-4

186. den Besten G, van Eunen K, Groen AK, Venema K, Reijngoud D-J, Bakker BM. The role of short-chain fatty acids in the interplay between diet, gut microbiota, and host energy metabolism. *J Lipid Res.* 2013;54(9). doi:10.1194/jlr.R036012

187. Primec M, Mičetić-Turk D, Langerholc T. Analysis of short-chain fatty acids in human feces: A scoping review. *Anal Biochem.* 2017;526:9–21. doi:10.1016/J.AB.2017.03.007

188. Sanna S, van Zuydam NR, Mahajan A, et al. Causal relationships among the gut microbiome, short-chain fatty acids and metabolic diseases. *Nat Genet.* 2019;51(4):600–605. doi:10.1038/s41588-019-0350-x

189. Scott AJ, Alexander JL, Merrifield CA, et al. International Cancer Microbiome Consortium consensus statement on the role of the human microbiome in carcinogenesis. *Gut.* 2019;68(9):1624–1632. doi:10.1136/gutjnl-2019-318556

190. TED Book: Follow Your Gut. https://www.ted.com/read/ted-books/ted-books-library/follow-your-gut. Accessed June 29, 2021.

191. Rothschild D, Weissbrod O, Barkan E, et al. Environment dominates over host genetics in shaping human gut microbiota. *Nature.* 2018;555(7695):210–215. doi:10.1038/nature25973

192. Xu F, Fu Y, Sun TY, et al. The interplay between host genetics and the gut microbiome reveals common and distinct microbiome features for complex human diseases. *Microbiome.* 2020;8(1). doi:10.1186/S40168-020-00923-9

193. Li J, Jia H, Cai X, et al. An integrated catalog of reference genes in the human gut microbiome. *Nat Biotechnol.* 2014;32(8):834–841. doi:10.1038/nbt.2942

194. Hildebrand F, Gossmann TI, Frioux C, et al. Dispersal strategies shape persistence and evolution of human gut bacteria. *Cell Host Microbe.* 2021;29(7):1167–1176.e9. doi:10.1016/J.CHOM.2021.05.008

195. Wu GD, Chen J, Hoffmann C, et al. Linking long-term dietary patterns with gut microbial enterotypes. *Science.* 2011;334(6052):105–108. doi:10.1126/science.1208344

196. Ley RE, Peterson DA, Gordon JI. Ecological and evolutionary forces shaping microbial diversity in the human intestine. *Cell.* 2006;124(4):837–848. doi:10.1016/J.CELL.2006.02.017

197. Ley RE, Turnbaugh PJ, Klein S, Gordon JI. Microbial ecology: Human gut microbes associated with obesity. *Nature.* 2006;444(7122):1022–1023. doi:10.1038/4441022A

198. David LA, Maurice CF, Carmody RN, et al. Diet rapidly and reproducibly alters the human gut microbiome. *Nature.* 2014;505(7484):559–563. doi:10.1038/nature12820

199. Weissman JL, Hou S, Fuhrman JA. Estimating maximal microbial growth rates from cultures, metagenomes, and single cells via codon usage patterns. *Proc Natl Acad Sci U S A.* 2021;118(12). doi:10.1073/PNAS.2016810118/-/DCSUPPLEMENTAL

200. Korem T, Zeevi D, Suez J, et al. Growth dynamics of gut microbiota in health and disease inferred from single metagenomic samples. *Science.* 2015;349(6252):1101. doi:10.1126/SCIENCE.AAC4812

201. Hansen NW, Sams A. The microbiotic highway to health— New perspective on food structure, gut microbiota, and host inflammation. *Nutrients.* 2018;10(11). doi:10.3390/nu10111590

202. Flint HJ, Scott KP, Duncan SH, Louis P, Forano E. Microbial degradation of complex carbohydrates in the gut. *Gut Microbes.* 2012;3(4):289. doi:10.4161/gmic.19897

203. Elemental. Why You Should Eat More Fiber. https://elemental. medium.com/if-you-really-want-to-optimize-your-diet-focus-on-fiber-c4ad231806f. Accessed July 7, 2021.

204. Tap J, Furet JP, Bensaada M, et al. Gut microbiota richness promotes its stability upon increased dietary fibre intake in healthy adults. *Environ Microbiol.* 2015;17(12):4954–4964. doi:10.1111/1462-2920.13006

205. Hald S, Schioldan AG, Moore ME, et al. Effects of arabinoxylan and resistant starch on intestinal microbiota and short-chain fatty acids in subjects with metabolic syndrome: A randomised crossover study. Loh G, ed. *PLoS One.* 2016;11(7):e0159223. doi:10.1371/journal.pone.0159223

206. Knudsen KEB, Lærke HN, Hedemann MS, et al. Impact of diet-modulated butyrate production on intestinal barrier function and inflammation. *Nutrients.* 2018;10(10). doi:10.3390/nu10101499

207. Overby HB, Ferguson JF. Gut microbiota-derived short-chain fatty acids facilitate microbiota: Host cross talk and modulate obesity and hypertension. *Curr Hypertens Rep.* 2021;23(2). doi:10.1007/s11906-020-01125-2

208. Ganda Mall JP, Löfvendahl L, Lindqvist CM, Brummer RJ, Keita V, Schoultz I. Differential effects of dietary fibres on colonic barrier function in elderly individuals with gastrointestinal symptoms. *Sci Rep.* 2018;8(1). doi:10.1038/s41598-018-31492-5

209. Valcheva R, Koleva P, Martínez I, Walter J, Gänzle MG, Dieleman LA. Insulin-type fructans improve active ulcerative colitis associated with microbiota changes and increased short-chain fatty acids levels. *Gut Microbes.* 2019;10(3):334–357. doi:10.1080/19490976.2018.1526583

210. de Vries J, Birkett A, Hulshof T, Verbeke K, Gibes K. Effects of cereal, fruit and vegetable fibers on human fecal weight and transit time: A comprehensive review of intervention trials. *Nutrients*. 2016;8(3):130. doi:10.3390/nu8030130

211. Quagliani D, Felt-Gunderson P. Closing America's fiber intake gap: Communication strategies from a food and fiber summit. *Am J Lifestyle Med*. 2017;11(1):80–85. doi:10.1177/1559827615588079

212. Lewis SJ, Heaton KW. The metabolic consequences of slow colonic transit. *Am J Gastroenterol*. 1999;94(8):2010–2016. doi:10.1111/j.1572-0241.1999.01271.x

213. Zeng H, Lazarova DL, Bordonaro M. Mechanisms linking dietary fiber, gut microbiota and colon cancer prevention. *World J Gastrointest Oncol*. 2014;6(2):41. doi:10.4251/WJGO.V6.I2.41

214. Chen S, Chen Y, Ma S, et al. Dietary fibre intake and risk of breast cancer: A systematic review and meta-analysis of epidemiological studies. *Oncotarget*. 2016;7(49):80980–80989. doi:10.18632/oncotarget.13140

215. O'Keefe SJ. The association between dietary fibre deficiency and high-income lifestyle-associated diseases: Burkitt's hypothesis revisited HHS public access. *Lancet Gastroenterol Hepatol*. 2019;4(12):984–996. doi:10.1016/S2468-1253(19)30257-2

216. Aune D, Chan DSM, Lau R, et al. Dietary fibre, whole grains, and risk of colorectal cancer: Systematic review and dose-response meta-analysis of prospective studies. *BMJ*. 2011;343(7833):1082. doi:10.1136/bmj.d6617

217. Bingham SA, Day NE, Luben R, et al. Dietary fibre in food and protection against colorectal cancer in the European

prospective investigation into cancer and nutrition (EPIC): An observational study. *Lancet.* 2003;361(9368):1496–1501. doi:10.1016/S0140-6736(03)13174-1

218. Morenga T, Reynolds A, Mann J, et al. Carbohydrate quality and human health: A series of systematic reviews and meta-analyses. *Lancet.* 2019;393(10170):434–445. doi:10.1016/S0140-6736(18)31809-9

219. Li D, Tong Y, Li Y. Dietary fiber is inversely associated with depressive symptoms in premenopausal women. *Front Neurosci.* 2020;14. doi:10.3389/fnins.2020.00373

220. Kim Y, Hong M, Kim S, Shin WY, Kim JH. Inverse association between dietary fiber intake and depression in premenopausal women: A nationwide population-based survey. *Menopause.* 2021;28(2):150–156. doi:10.1097/GME.0000000000001711

221. Hovey MHN AL, Jones GP, Devereux HM, Walker MND KZ. *Whole Cereal and Legume Seeds Increase Faecal Short Chain Fatty Acids Compared to Ground Seeds.* Vol 12; 2003.

222. Tullio V, Gasperi V, Catani MV, Savini I. The impact of whole grain intake on gastrointestinal tumors: A focus on colorectal, gastric and esophageal cancers. *Nutrients.* 2021;13(1):1–25. doi:10.3390/nu13010081

223. Burkitt DP, Walker ARP, Painter NS. Effect of dietary fibre on stools and the transit-times, and its role in the causation of disease. *Lancet (London, England).* 1972;2(7792):1408–1411. doi:10.1016/S0140-6736(72)92974-1

224. Burkitt DP. An epidemiologic approach to cancer of the large intestine: The significance of disease relationships. *Dis Colon Rectum.* 1974;17(4):456–461. doi:10.1007/BF02587020

225. D P Burkitt HCT. Dietary fibre and western diseases. https:// pubmed.ncbi.nlm.nih.gov/893060/. Published 1977. Accessed March 1, 2022.

226. Sonnenburg ED, Sonnenburg JL. Starving our microbial self: The deleterious consequences of a diet deficient in microbiota-accessible carbohydrates. *Cell Metab.* 2014;20(5):779–786. doi:10.1016/j.cmet.2014.07.003

227. Martens EC. Microbiome: Fibre for the future. *Nature.* 2016;529(7585):158–159. doi:10.1038/529158a

228. Duncan SH, Belenguer A, Holtrop G, Johnstone AM, Flint HJ, Lobley GE. Reduced dietary intake of carbohydrates by obese subjects results in decreased concentrations of butyrate and butyrate-producing bacteria in feces. *Appl Environ Microbiol.* 2007;73(4):1073–1078. doi:10.1128/AEM.02340-06

229. Brinkworth GD, Noakes M, Clifton PM, Bird AR. Comparative effects of very low-carbohydrate, high-fat and high-carbohydrate, low-fat weight-loss diets on bowel habit and faecal short-chain fatty acids and bacterial populations. *Br J Nutr.* 2009;101(10):1493–1502. doi:10.1017/S0007114508094658

230. Brinkworth GD, Noakes M, Buckley JD, Keogh JB, Clifton PM. Long-term effects of a very-low-carbohydrate weight loss diet compared with an isocaloric low-fat diet after 12 mo. *Am J Clin Nutr.* 2009;90(1):23–32. doi:10.3945/ajcn.2008.27326

231. Desai MS, Seekatz AM, Koropatkin NM, et al. A dietary fiber-deprived gut Microbiota degrades the colonic mucus barrier and enhances pathogen susceptibility. *Cell.* 2016;167(5): 1339–1353.e21. doi:10.1016/j.cell.2016.10.043

232. Krack A, Sharma R, Figulla HR, Anker SD. The importance of the gastrointestinal system in the pathogenesis of heart

failure. *Eur Heart J.* 2005;26(22):2368–2374. doi:10.1093/
EURHEARTJ/EHI389

233. Venegas DP, De La Fuente MK, Landskron G, et al. Short
chain fatty acids (SCFAs)mediated gut epithelial and
immune regulation and its relevance for inflammatory bowel
diseases. *Front Immunol.* 2019;10(MAR):277. doi:10.3389/
fimmu.2019.00277

234. Jakobsson HE, Rodríguez-Piñeiro AM, Schütte A, et al.
The composition of the gut microbiota shapes the colon
mucus barrier. *EMBO Rep.* 2015;16(2):164–177. doi:10.15252/
embr.201439263

235. Johansson, M. Gustafsson, J. Holmen-Larsson, J. et al. Bacteria
penetrate the normally impenetrable inner colon mucus layer in
both murine colitis models and patients with ulcerative colitis.
Gut. 2014;63(2):281–291. doi:10.1136/GUTJNL-2012-303207

236. Vipperla K, O'Keefe SJ. Diet, microbiota, and dysbiosis: A
"recipe" for colorectal cancer. *Food Funct.* 2016;7(4):1731–1740.
doi:10.1039/c5fo01276g

237. DGA. Dietary Guidelines for Americans, 2020–2025 and
Online Materials. https://www.dietaryguidelines.gov/
resources/2020-2025-dietary-guidelines-online-materials.
Accessed March 2, 2022.

238. Foundation P. 2020 PBH State of the Plate Executive
Summary—Have a Plant. https://fruitsandveggies.org/
stateoftheplate2020/. Published 2020. Accessed March 2, 2022.

239. CDC. NHANES—What We Eat in America. https://www.cdc.
gov/nchs/nhanes/wweia.htm. Accessed March 2, 2022.

240. DGA. Food Sources of Select Nutrients. https://www.
dietaryguidelines.gov/

resources/2020-2025-dietary-guidelines-online-materials/
food-sources-select-nutrients. Accessed March 3, 2022.

241. Ma Y, Hu M, Zhou L, et al. Dietary fiber intake and risks of
proximal and distal colon cancers: A meta-analysis. *Med
(United States)*. 2018;97(36). doi:10.1097/MD.0000000000011678

242. Histologically diagnosed cancers in South Africa, 1988.
PubMed. https://pubmed.ncbi.nlm.nih.gov/7740381/. Accessed
January 13, 2022.

243. O'Keefe SJD. Rarity of colon cancer in Africans is associated
with low animal product consumption, not fiber. *Am J
Gastroenterol*. 1999;94(5):1373–1380. doi:10.1111/j.1572-0241.
1999.01089.x

244. Mensah DO, Nunes AR, Bockarie T, Lillywhite R, Oyebode O.
Meat, fruit, and vegetable consumption in sub-Saharan Africa:
A systematic review and meta-regression analysis. *Nutr Rev*.
2021;79(6):651–692. doi:10.1093/NUTRIT/NUAA032

245. Bashir S, Fezeu LK, Ben-Arye SL, et al. Association between
Neu5Gc carbohydrate and serum antibodies against it
provides the molecular link to cancer: French NutriNet-
Santé study. *BMC Med 2020 181*. 2020;18(1):1–19. doi:10.1186/
S12916-020-01721-8

246. WHO. Cancer: Carcinogenicity of the consumption of red
meat and processed meat. https://www.who.int/news-room/
q-a-detail/cancer-carcinogenicity-of-the-consumption-of-red-
meat-and-processed-meat. Accessed June 5, 2021.

247. Domingo JL, Nadal M. Carcinogenicity of consumption of
red meat and processed meat: A review of scientific news
since the IARC decision. *Food Chem Toxicol*. 2017;105:256–261.
doi:10.1016/J.FCT.2017.04.028

248. Smith T. Some problems in the life-history of pathogenic microorganisms. *Amer Med.* 1904;viii:711–718.

249. Kendall AI. *Some observations on the study of the intestinal bacteria.* http://www.jbc.org/. Accessed March 31, 2020.

250. Diether NE, Willing BP. Microbial fermentation of dietary protein: An important factor in diet–microbe–host interaction. *Microorganisms.* 2019;7(1). doi:10.3390/MICROORGANISMS7010019

251. Oliphant K, Allen-Vercoe E. Macronutrient metabolism by the human gut microbiome: Major fermentation by-products and their impact on host health. *Microbiome.* 2019;7(1):1–15. doi:10.1186/s40168-019-0704-8

252. Duda-Chodak A, Tarko T, Satora P, Sroka P. Interaction of dietary compounds, especially polyphenols, with the intestinal microbiota: A review. *Eur J Nutr.* 2015;54(3):325. doi:10.1007/S00394-015-0852-Y

253. A sulfosugar from green vegetables promotes the growth of important gut bacteria. *ScienceDaily.* https://www.sciencedaily.com/releases/2021/04/210409104447.htm?utm_source=feedburner&utm_medium=email&utm_campaign=Feed%3A+sciencedaily%2Fhealth_medicine%2Fnutrition+%28Nutrition+News+–+ScienceDaily%29. Accessed July 22, 2021.

254. Frassetto L. Sulfur-containing amino acid content of common foods. https://www.researchgate.net/figure/Sulfur-Containing-Amino-Acid-Content-of-Common-Foods-of-Vegetable-and-Animal-Origin_tbl2_12290625. Published 2000. Accessed January 19, 2022.

255. Hansen JM, Jones DP. Thiols in cancer. *Nutr Oncol.* January 2006:307–320. doi:10.1016/B978-012088393-6/50071-3

256. Wallace JL, Motta JP, Buret AG. Hydrogen sulfide: An agent of stability at the microbiome-mucosa interface. *Am J Physiol - Gastrointest Liver Physiol.* 2018;314(2):G143–G149. doi:10.1152/ajpgi.00249.2017

257. Zanardo RCO, Brancaleone V, Distrutti E, et al. Hydrogen sulfide is an endogenous modulator of leukocyte-mediated inflammation. *FASEB J.* 2006;20(12):2118–2120. doi:10.1096/FJ.06-6270FJE

258. Leschelle X, Goubern M, Andriamihaja M, et al. Adaptative metabolic response of human colonic epithelial cells to the adverse effects of the luminal compound sulfide. *Biochim Biophys Acta.* 2005;1725(2):201–212. doi:10.1016/J.BBAGEN.2005.06.002

259. Dong Z, Gao X, Chinchilli VM, et al. Association of sulfur amino acid consumption with cardiometabolic risk factors: Cross-sectional findings from NHANES III. *EClinicalMedicine.* 2020;19. doi:10.1016/J.ECLINM.2019.100248

260. David LA, Maurice CF, Carmody RN, et al. Diet rapidly and reproducibly alters the human gut microbiome. *Nature.* 2014;505(7484). doi:10.1038/nature12820

261. Magee EA, Richardson CJ, Hughes R, Cummings JH. Contribution of dietary protein to sulfide production in the large intestine: An in vitro and a controlled feeding study in humans. *Am J Clin Nutr.* 2000;72(6):1488–1494. doi:10.1093/AJCN/72.6.1488

262. Ijssennagger N, Belzer C, Hooiveld GJ, et al. Gut microbiota facilitates dietary heme-induced epithelial hyperproliferation by opening the mucus barrier in colon. *Proc Natl Acad Sci U S A.* 2015;112(32):10038–10043. doi:10.1073/pnas.1507645112

263. Ijssennagger N, van der Meer R, van Mil SWC. Sulfide as a mucus barrier-breaker in inflammatory bowel disease? *Trends Mol Med.* 2016;22(3):190–199. doi:10.1016/J.MOLMED. 2016.01.002

264. Attene-Ramos MS, Wagner ED, Plewa MJ, Gaskins HR. Evidence that hydrogen sulfide is a genotoxic agent. *Mol Cancer Res.* 2006;4(1):9–14. doi:10.1158/1541-7786.MCR-05-0126

265. Attene-Ramos MS, Wagner ED, Gaskins HR, Plewa MJ. Hydrogen sulfide induces direct radical-associated DNA damage. *Mol Cancer Res.* 2007;5(5):455–459. doi:10.1158/1541-7786.MCR-06-0439

266. Szabo C, Coletta C, Chao C, et al. Tumor-derived hydrogen sulfide, produced by cystathionine-β-synthase, stimulates bioenergetics, cell proliferation, and angiogenesis in colon cancer. *Proc Natl Acad Sci U S A.* 2013;110(30):12474–12479. doi:10.1073/PNAS.1306241110

267. Zeller G, Tap J, Voigt AY, et al. Potential of fecal microbiota for early-stage detection of colorectal cancer. *Mol Syst Biol.* 2014;10(11):766. doi:10.15252/msb.20145645

268. Dordević D, Jančíková S, Vítězová M, Kushkevych I. Hydrogen sulfide toxicity in the gut environment: Meta-analysis of sulfate-reducing and lactic acid bacteria in inflammatory processes. *J Adv Res.* 2021;27:55–69. doi:10.1016/j.jare. 2020.03.003

269. Levine J, Ellis CJ, Furne JK, Springfield J, Levitt MD. Fecal hydrogen sulfide production in ulcerative colitis. *Am J Gastroenterol.* 1998;93(1):83–87. doi:10.1111/J.1572-0241.1998.083_C.X

270. Carbonero F, Benefiel AC, Alizadeh-Ghamsari AH, Gaskins HR. Microbial pathways in colonic sulfur metabolism and

links with health and disease. *Front Physiol.* 2012;3 NOV:448. doi:10.3389/fphys.2012.00448

271. Hou JK, Abraham B, El-Serag H. Dietary intake and risk of developing inflammatory bowel disease: A systematic review of the literature. *Am J Gastroenterol.* 2011;106(4):563–573. doi:10.1038/ajg.2011.44

272. Truelove SC. Ulcerative colitis provoked by milk. *Br Med J.* 1961;1(5220):154. doi:10.1136/BMJ.1.5220.154

273. Chiba M, Abe T, Tsuda H, et al. Lifestyle-related disease in Crohn's disease: Relapse prevention by a semi-vegetarian diet. 2019. doi:10.3748/wjg.v16.i20.2484

274. Medani M, Collins D, Docherty NG, Baird AW, O'Connell PR, Winter DC. Emerging role of hydrogen sulfide in colonic physiology and pathophysiology. *Inflamm Bowel Dis.* 2011;17(7):1620–1625. doi:10.1002/IBD.21528

275. Carco C, Young W, Gearry RB, Talley NJ, McNabb WC, Roy NC. Increasing evidence that irritable bowel syndrome and functional gastrointestinal disorders have a microbial pathogenesis. *Front Cell Infect Microbiol.* 2020;10:468. doi:10.3389/FCIMB.2020.00468

276. Sanctuary MR, Kain JN, Angkustsiri K, German JB. Dietary considerations in autism spectrum disorders: The potential role of protein digestion and microbial putrefaction in the gut-brain axis. *Front Nutr.* 2018;5. doi:10.3389/FNUT.2018.00040

277. Jimenez M, Gil V, Martinez-Cutillas M, Mañé N, Gallego D. Hydrogen sulphide as a signalling molecule regulating physiopathological processes in gastrointestinal motility. *Br J Pharmacol.* 2017;174(17):2805. doi:10.1111/BPH.13918

278. Babidge W, Millard S, Roediger W. Sulfides impair short chain fatty acid β-oxidation at acyl-CoA dehydrogenase level in colonocytes: Implications for ulcerative colitis. *Mol Cell Biochem 1998 1811.* 1998;181(1):117–124. doi:10.1023/A:1006838231432

279. Birkett A, Muir J, Phillips J, Jones G, O'Dea K. Resistant starch lowers fecal concentrations of ammonia and phenols in humans. *Am J Clin Nutr.* 1996;63(5):766–772. doi:10.1093/AJCN/63.5.766

280. Hylla S, Gostner A, Dusel G, et al. Effects of resistant starch on the colon in healthy volunteers: Possible implications for cancer prevention. *Am J Clin Nutr.* 1998;67(1):136–142. doi:10.1093/ajcn/67.1.136

281. Dong Z, Gao X, Chinchilli VM, et al. Association of sulfur amino acid consumption with cardiometabolic risk factors: Cross-sectional findings from NHANES III. *EClinicalMedicine.* 2020;19. doi:10.1016/J.ECLINM.2019.100248

282. Abid MA, Abid MB. Commentary: Dietary methionine influences therapy in mouse cancer models and alters human metabolism. *Front Oncol.* 2020;10:1071. doi:10.3389/fonc.2020.01071

283. Gao X, Sanderson SM, Dai Z, et al. Dietary methionine influences therapy in mouse cancer models and alters human metabolism. *Nature.* 2019;572(7769):397–401. doi:10.1038/s41586-019-1437-3

284. Durando X, Farges MC, Buc E, et al. Dietary methionine restriction with FOLFOX regimen as first line therapy of metastatic colorectal cancer: A feasibility study. *Oncology.* 2010;78(3-4):205–209. doi:10.1159/000313700

285. Yoo W, Zieba JK, Foegeding NJ, et al. High-fat diet-induced colonocyte dysfunction escalates microbiota-derived trimethylamine n-oxide. *Science.* 2021;373(6556):813–818. doi:10.1126/SCIENCE.ABA3683

286. Armstrong H, Bording-Jorgensen M, Wine E. The multifaceted roles of diet, microbes, and metabolites in cancer. *Cancers (Basel)*. 2021;13(4):767. doi:10.3390/cancers13040767

287. Cani PD, Amar J, Iglesias MA, et al. Metabolic endotoxemia initiates obesity and insulin resistance. *Diabetes*. 2007;56(7):1761–1772. doi:10.2337/DB06-1491

288. Harte AL, Varma MC, Tripathi G, et al. High fat intake leads to acute postprandial exposure to circulating endotoxin in type 2 diabetic subjects. *Diabetes Care*. 2012;35(2):375–382. doi:10.2337/dc11-1593

289. Wan Y, Wang F, Yuan J, et al. Effects of dietary fat on gut microbiota and faecal metabolites, and their relationship with cardiometabolic risk factors: A 6-month randomised controlled-feeding trial. *Gut*. 2019;68(8):1417–1429. doi:10.1136/gutjnl-2018-317609

290. Erridge C, Attina T, Spickett CM, Webb DJ. A high-fat meal induces low-grade endotoxemia: Evidence of a novel mechanism of postprandial inflammation. *Am J Clin Nutr*. 2007;86(5):1286–1292. doi:10.1093/ajcn/86.5.1286

291. Reddy BS, Weisburger JH, Wynder EL. Fecal bacterial β-glucuronidase: Control by diet. *Science*. 1974;183(4123): 416–417. doi:10.1126/SCIENCE.183.4123.416

292. Chamseddine AN, Ducreux M, Armand JP, et al. Intestinal bacterial β-glucuronidase as a possible predictive biomarker of irinotecan-induced diarrhea severity. *Pharmacol Ther*. 2019;199:1–15. doi:10.1016/j.pharmthera.2019.03.002

293. Kim M, Vogtmann E, Ahlquist DA, et al. Fecal metabolomic signatures in colorectal adenoma patients are associated with gut microbiota and early events of colorectal cancer pathogenesis. *MBio*. 2020;11(1). doi:10.1128/mBio.03186-19

294. Sperker B, Werner U, Murdter TE, et al. Expression and function of beta-glucuronidase in pancreatic cancer: Potential role in drug targeting. *Naunyn Schmiedebergs Arch Pharmacol.* 2000;362(2):110–115. doi:10.1007/S002100000260

295. Kwa M, Plottel CS, Blaser MJ, Adams S. The intestinal microbiome and estrogen receptor-positive female breast cancer. *J Natl Cancer Inst.* 2016;108(8). doi:10.1093/JNCI/DJW029

296. Bhatt AP, Pellock SJ, Biernat KA, et al. Targeted inhibition of gut bacterial β-glucuronidase activity enhances anticancer drug efficacy. *Proc Natl Acad Sci U S A.* 2020;117(13):7374–7381. doi:10.1073/PNAS.1918095117

297. Ling WH, Hänninen O. Shifting from a conventional diet to an uncooked vegan diet reversibly alters fecal hydrolytic activities in humans. *J Nutr.* 1992;122(4):924–930. doi:10.1093/jn/122.4.924

298. Study reveals missing link between high-fat diet, microbiota and heart disease. *ScienceDaily.* https://www.sciencedaily.com/releases/2021/08/210812145052.htm?utm_source=feedburner&utm_medium=email&utm_campaign=Feed%3A+sciencedaily%2Fhealth_medicine%2Fnutrition+%28Nutrition+News+–+ScienceDaily%29. Accessed November 19, 2021.

299. Zeng H, Umar S, Rust B, Lazarova D, Bordonaro M. Secondary bile acids and short chain fatty acids in the colon: A focus on colonic microbiome, cell proliferation, inflammation, and cancer. *Int J Mol Sci.* 2019;20(5). doi:10.3390/ijms20051214

300. Farhana L, Nangia-Makker P, Arbit E, et al. Bile acid: A potential inducer of colon cancer stem cells. *Stem Cell Res Ther.* 2016;7(1). doi:10.1186/s13287-016-0439-4

301. Chiang JYL. Bile acids: Regulation of synthesis. *J Lipid Res.* 2009;50(10):1955–1966. doi:10.1194/JLR.R900010-JLR200

302. Hang S, Paik D, Yao L, et al. Bile acid metabolites control
     TH17 and Treg cell differentiation. *Nat 2019 5767785*.
     2019;576(7785):143–148. doi:10.1038/s41586-019-1785-z

303. Murakami Y, Tanabe S, Suzuki T. High-fat diet-induced
     intestinal hyperpermeability is associated with increased
     bile acids in the large intestine of mice. *J Food Sci.*
     2016;81(1):H216–H222. doi:10.1111/1750-3841.13166

304. Bernstein C et al. A bile acid-induced apoptosis assay for
     colon cancer risk and associated quality control studies.
     https://pubmed.ncbi.nlm.nih.gov/10344743/. Published 1999.
     Accessed March 7, 2022.

305. Ridlon JM, Kang DJ, Hylemon PB. Bile salt biotransformations
     by human intestinal bacteria. *J Lipid Res.* 2006;47(2):241–259.
     doi:10.1194/JLR.R500013-JLR200

306. Ajouz H, Mukherji D, Shamseddine A. Secondary bile acids:
     An underrecognized cause of colon cancer. *World J Surg Oncol.*
     2014;12(1). doi:10.1186/1477-7819-12-164

307. Pratt M, Forbes JD, Knox NC, Bernstein CN, Van Domselaar G.
     Microbiome-mediated immune signaling in inflammatory
     bowel disease and colorectal cancer: Support from meta-
     omics data. *Front Cell Dev Biol.* 2021;9:716604. doi:10.3389/
     fcell.2021.716604

308. Cook JW, Kennaway EL, Kennaway NM. Production of
     tumours in mice by deoxycholic acid. *Nat 1940 1453677*.
     1940;145(3677):627–627. doi:10.1038/145627a0

309. Stenman LK, Holma R, Korpcla R. High-fat-induced intestinal
     permeability dysfunction associated with altered fecal bile
     acids. *World J Gastroenterol.* 2012;18(9):923. doi:10.3748/WJG.
     V18.I9.923

310. Dermadi D, Valo S, Ollila S, et al. Western diet deregulates bile acid homeostasis, cell proliferation, and tumorigenesis in colon. *Cancer Res.* 2017;77(12):3352–3363. doi:10.1158/0008-5472.CAN-16-2860

311. Payne CM, Bernstein C, Dvorak K, Bernstein H. Hydrophobic bile acids, genomic instability, Darwinian selection, and colon carcinogenesis. *Clin Exp Gastroenterol.* 2008;1:19. doi:10.2147/CEG.S4343

312. Yoshimoto S, Loo TM, Atarashi K, et al. Obesity-induced gut microbial metabolite promotes liver cancer through senescence secretome. *Nature.* 2013;499(7456):97–101. doi:10.1038/NATURE12347

313. Gupta B, Liu Y, Chopyk DM, et al. Western diet-induced increase in colonic bile acids compromises epithelial barrier in nonalcoholic steatohepatitis. *FASEB J.* 2020;34(5):7089–7102. doi:10.1096/fj.201902687R

314. Wu L, Feng J, Li J, et al. The gut microbiome-bile acid axis in hepatocarcinogenesis. *Biomed Pharmacother.* 2021;133. doi:10.1016/J.BIOPHA.2020.111036

315. Costarelli V, Sanders TAB. Plasma deoxycholic acid concentration is elevated in postmenopausal women with newly diagnosed breast cancer. *Eur J Clin Nutr.* 2002;56(9):925–927. doi:10.1038/sj.ejcn.1601396

316. Alok A, Lei Z, Jagannathan NS, et al. Wnt proteins synergize to activate β-catenin signaling. *J Cell Sci.* 2017;130(9):1532–1544. doi:10.1242/JCS.198093

317. Nusse R, Clevers H. Wnt/β-catenin signaling, disease, and emerging therapeutic modalities. *Cell.* 2017;169(6):985–999. doi:10.1016/J.CELL.2017.05.016

318. Bayerdörffer E, Mannes GA, Richter WO, et al. Increased serum deoxycholic acid levels in men with colorectal adenomas. *Gastroenterology.* 1993;104(1):145–151. doi:10.1016/0016-5085(93)90846-5

319. Bayerdörffer E, Mannes GA, Ochsenkühn T, Dirschedl P, Wiebecke B, Paumgartner G. Unconjugated secondary bile acids in the serum of patients with colorectal adenomas. *Gut.* 1995;36(2):268–273. doi:10.1136/gut.36.2.268

320. Javitt NB, Budai K, Raju U, Levitz M, Miller DG, Cahan AC. Breast-gut connection: Origin of chenodeoxycholic acid in breast cyst fluid. *Lancet.* 1994;343(8898):633–635. doi:10.1016/S0140-6736(94)92635-2

321. Raju U, Levitz M, Javitt NB. Bile acids in human breast cyst fluid: The identification of lithocholic acid. *J Clin Endocrinol Metab.* 1990;70(4):1030–1034. doi:10.1210/JCEM-70-4-1030

322. Journe F, Durbecq V, Chaboteaux C, et al. Association between farnesoid X receptor expression and cell proliferation in estrogen receptor-positive luminal-like breast cancer from postmenopausal patients. *Breast Cancer Res Treat.* 2009;115(3):523–535. doi:10.1007/S10549-008-0094-2

323. Wang DD, Nguyen LH, Li Y, et al. The gut microbiome modulates the protective association between a Mediterranean diet and cardiometabolic disease risk. *Nat Med.* 2021;27(2):333–343. doi:10.1038/s41591-020-01223-3

324. Estruch R, Ros E, Salas-Salvadó J, et al. Primary prevention of cardiovascular disease with a Mediterranean diet supplemented with extra-virgin olive oil or nuts. *N Engl J Med.* 2018;378(25):e34. doi:10.1056/NEJMOA1800389

325. Toledo E, Salas-Salvado J, Donat-Vargas C, et al. Mediterranean diet and invasive breast cancer risk among women at high cardiovascular risk in the predimed trial a randomized clinical trial. *JAMA Intern Med.* 2015;175(11): 1752–1760. doi:10.1001/jamainternmed.2015.4838

326. Wan Y, Wu K, Wang L, et al. Dietary fat and fatty acids in relation to risk of colorectal cancer. *Eur J Nutr 2021.* January 2022:1–11. doi:10.1007/S00394-021-02777-9

327. Lombardi-Boccia G, Martinez-Dominguez B, Aguzzi A. Total heme and non-heme iron in raw and cooked meats. *J Food Sci.* 2002;67(5):1738–1741. doi:10.1111/J.1365-2621.2002.TB08715.X

328. Seiwert N, Heylmann D, Hasselwander S, Fahrer J. Mechanism of colorectal carcinogenesis triggered by heme iron from red meat. *Biochim Biophys Acta Rev Cancer.* 2020;1873(1). doi:10.1016/J.BBCAN.2019.188334

329. Glei M, Klenow S, Sauer J, Wegewitz U, Richter K, Pool-Zobel BL. Hemoglobin and hemin induce DNA damage in human colon tumor cells HT29 clone 19A and in primary human colonocytes. *Mutat Res.* 2006;594(1-2):162–171. doi:10.1016/ J.MRFMMM.2005.08.006

330. Constante M, Fragoso G, Calvé A, Samba-Mondonga M, Santos MM. Dietary heme induces gut dysbiosis, aggravates colitis, and potentiates the development of adenomas in mice. *Front Microbiol.* 2017;8(SEP):1809. doi:10.3389/ fmicb.2017.01809

331. Seiwert N, Wecklein S, Demuth P, et al. Heme oxygenase 1 protects human colonocytes against ROS formation, oxidative DNA damage and cytotoxicity induced by heme iron, but not inorganic iron. *Cell Death Dis 2020 119.* 2020;11(9):1–16. doi:10.1038/s41419-020-02950-8

332. Cross AJ, Harnly JM, Ferrucci LM, Risch A, Mayne ST, Sinha R. Developing a heme iron database for meats according to meat type, cooking method and doneness level. *Food Nutr Sci.* 2012;3(7):905. doi:10.4236/FNS.2012.37120

333. Schwartz S, Ellefson M. Quantitative fecal recovery of ingested hemoglobin-heme in blood: Comparisons by HemoQuant assay with ingested meat and fish. *Gastroenterology.* 1985;89(1):19–26. doi:10.1016/0016-5085(85)90740-1

334. Bouvard V, Loomis D, Guyton KZ, et al. Carcinogenicity of consumption of red and processed meat. *Lancet Oncol.* 2015;16(16):1599–1600. doi:10.1016/S1470-2045(15)00444-1

335. Bastide NM, Pierre FHF, Corpet DE. Heme iron from meat and risk of colorectal cancer: A meta-analysis and a review of the mechanisms involved. *Cancer Prev Res.* 2011;4(2):177–184. doi:10.1158/1940-6207.CAPR-10-0113

336. Kabat GC, Miller AB, Jain M, Rohan TE. A cohort study of dietary iron and heme iron intake and risk of colorectal cancer in women. *Br J Cancer.* 2007;97(1):118–122. doi:10.1038/SJ.BJC.6603837

337. Cross AJ, Ferrucci LM, Risch A, et al. A large prospective study of meat consumption and colorectal cancer risk: An investigation of potential mechanisms underlying this association. *Cancer Res.* 2010;70(6):2406–2414. doi:10.1158/0008-5472.CAN-09-3929

338. Bastide N, Morois S, Cadeau C, et al. Heme iron intake, dietary antioxidant capacity, and risk of colorectal adenomas in a large cohort study of French women. *Cancer Epidemiol Biomarkers Prev.* 2016;25(4):640–647. doi:10.1158/1055-9965.EPI-15-0724

339. McCullough ML, Hodge RA, Campbell PT, Stevens VL, Wang Y. Pre-diagnostic circulating metabolites and colorectal cancer risk in the cancer prevention study-II nutrition cohort. *Metabolites.* 2021;11(3). doi:10.3390/metabo11030156

340. Le Leu RK, Winter JM, Christophersen CT, et al. Butyrylated starch intake can prevent red meat-induced O6-methyl-2-deoxyguanosine adducts in human rectal tissue: A randomised clinical trial. *Br J Nutr.* 2015;114(2):220–230. doi:10.1017/S0007114515001750

341. Lewin MH, Bailey N, Bandaletova T, et al. Red meat enhances the colonic formation of the DNA adduct o 6-carboxymethyl guanine: Implications for colorectal cancer risk. *Cancer Res.* 2006;66(3):1859–1865. doi:10.1158/0008-5472.CAN-05-2237

342. Kuhnle GG, Story GW, Reda T, et al. Diet-induced endogenous formation of nitroso compounds in the GI tract. *Free Radic Biol Med.* 2007;43(7):1040–1047. doi:10.1016/J.FREERADBIOMED.2007.03.011

343. Bruce W. A mutagen in the feces of normal humans. *Orig Hum Cancer.* 1977:1641–1646.

344. Song P, Wu L, Guan W. Dietary nitrates, nitrites, and nitrosamines intake and the risk of gastric cancer: A meta-analysis. *Nutrients.* 2015;7(12):9872–9895. doi:10.3390/nu7125505

345. Hughes R, Cross AJ, Pollock JR, Bingham S. Dose-dependent effect of dietary meat on endogenous colonic N-nitrosation. *Carcinogenesis.* 2001;22(1):199–202. doi:10.1093/carcin/22.1.199

346. Huncharek M, Kupelnick B. A meta-analysis of maternal cured meat consumption during pregnancy and the risk of childhood brain tumors. *Neuroepidemiology.* 2004;23(1-2):78–84. doi:10.1159/000073979

347. Huncharek M, Kupelnick B, Wheeler L. Dietary cured meat and the risk of adult glioma: A meta-analysis of nine observational studies. *J Environ Pathol Toxicol Oncol.* 2003;22(2):129–137. doi:10.1615/JEnvPathToxOncol. v22.i2.60

348. Blot WJ, Henderson BE, Boice JD. Childhood cancer in relation to cured meat intake: Review of the epidemiological evidence. *Nutr Cancer.* 1999;34(1):111–118. doi:10.1207/S15327914NC340115

349. Issenberg P. Nitrites, nitrosamines, and cancer. *Lancet.* 1968;291(7551):1071–1072. doi:10.1016/S0140-6736(68)91418-9

350. Maffei F, Moraga JMZ, Angelini S, et al. Micronucleus frequency in human peripheral blood lymphocytes as a biomarker for the early detection of colorectal cancer risk. *Mutagenesis.* 2014;29(3):221–225. doi:10.1093/MUTAGE/GEU007

351. Hebels D, Kok D, Van Herwijnen MC, et al. Red meat intake-induced increases in fecal water genotoxicity correlate with pro-carcinogenic gene expression changes in the human colon. 2011. doi:10.1016/j.fct.2011.10.038

352. Gratz SW, Wallace RJ, El-Nezami HS. Recent perspectives on the relations between fecal mutagenicity, genotoxicity, and diet. *Front Pharmacol.* 2011;MAR. doi:10.3389/fphar.2011.00004

353. Media Centre – IARC News. https://www.iarc.who.int/media-centre-iarc-news-49/. Accessed January 26, 2022.

354. Mirvish SS. Effects of vitamins c and e on carcinogen formation and action, and relationship to human cancer. *Basic Life Sci.* 1986;39:83–85. doi:10.1007/978-1-4684-5182-5_7

355. Dellavalle CT, Xiao Q, Yang G, et al. Dietary nitrate and nitrite intake and risk of colorectal cancer in the Shanghai Women's Health Study. *Int J Cancer.* 2014;134(12):2917–2926. doi:10.1002/IJC.28612

356. Lunn JC, Kuhnle G, Mai V, et al. The effect of haem in red and processed meat on the endogenous formation of N-nitroso compounds in the upper gastrointestinal tract. *Carcinogenesis.* 2007;28(3):685–690. doi:10.1093/CARCIN/BGL192

357. Joosen AMCP, Kuhnle GGC, Aspinall SM, et al. Effect of processed and red meat on endogenous nitrosation and DNA damage. *Carcinogenesis.* 2009;30(8):1402–1407. doi:10.1093/CARCIN/BGP130

358. Massey R. An investigation of the endogenous formation of apparent total N-nitroso compounds in conventional flora and germ-free rats. *Food Chem Toxicol.* 1991;(26):595–600.

359. Kobayashi J. Effect of diet and gut environment on the gastrointestinal formation of N-nitroso compounds: A review. *Nitric Oxide Biol Chem.* 2018;73:66–73. doi:10.1016/J.NIOX.2017.06.001

360. Hemeryck LY, Rombouts C, De Paepe E, Vanhaecke L. DNA adduct profiling of in vitro colonic meat digests to map red vs. white meat genotoxicity. *Food Chem Toxicol.* 2018;115:73–87. doi:10.1016/J.FCT.2018.02.032

361. Martin OCB, Lin C, Naud N, et al. Antibiotic suppression of intestinal microbiota reduces heme-induced lipoperoxidation associated with colon carcinogenesis in rats. *Nutr Cancer.* 2015;67(1):119–125. doi:10.1080/01635581.2015.976317

362. Rak K, Rader DJ. Cardiovascular disease: The diet-microbe morbid union. *Nature.* 2011;472(7341):40–41. doi:10.1038/472040a

363. Tang WH, Wang Z, Levison BS, et al. Intestinal microbial metabolism of phosphatidylcholine and cardiovascular risk. *N Engl J Med.* 2013;368(17):1575–1584. doi:10.1056/NEJMOA1109400

364. Koeth RA, Wang Z, Levison BS, et al. Intestinal microbiota metabolism of L-carnitine, a nutrient in red meat, promotes atherosclerosis. *Nat Med.* 2013;19(5):576–585. doi:10.1038/nm.3145

365. Qi J, You T, Li J, et al. Circulating trimethylamine n-oxide and the risk of cardiovascular diseases: A systematic review and meta-analysis of 11 prospective cohort studies. *J Cell Mol Med.* 2018;22(1):185–194. doi:10.1111/jcmm.13307

366. Wang Z, Klipfell E, Bennett BJ, et al. Gut flora metabolism of phosphatidylcholine promotes cardiovascular disease. *Nat.* 2011;472(7341):57–63. doi:10.1038/nature09922

367. Mente A, Chalcraft K, Ak H, et al. The relationship between trimethylamine N-oxide and prevalent cardiovascular disease in a multiethnic population living in Canada. *Can J Cardiol.* 2015;31(9):1189–1194. doi:10.1016/J.CJCA.2015.06.016

368. Tang WHW, Wang Z, Levison BS, et al. Intestinal microbial metabolism of phosphatidylcholine and cardiovascular risk. *N Engl J Med.* 2013;368(17):1575–1584. doi:10.1056/nejmoa1109400

369. Koeth RA, Lam-Galvez BR, Kirsop J, et al. L-carnitine in omnivorous diets induces an atherogenic gut microbial pathway in humans. *J Clin Invest.* 2019;129(1). doi:10.1172/JCI94601

370. Taesuwan S, Cho CE, Malysheva O V., et al. The metabolic fate of isotopically labeled trimethylamine-N-oxide (TMAO) in humans. *J Nutr Biochem.* 2017;45:77–82. doi:10.1016/J.JNUTBIO.2017.02.010

371. Genoni A, Christophersen CT, Lo J, et al. Long-term Paleolithic diet is associated with lower resistant starch intake, different gut microbiota composition and increased serum TMAO concentrations. *Eur J Nutr.* 2020;59(5):1845–1858. doi:10.1007/s00394-019-02036-y

372. Genoni A, Christophersen CT, Lo J, et al. Long-term Paleolithic diet is associated with lower resistant starch intake, different gut microbiota composition and increased serum TMAO concentrations. *Eur J Nutr.* July 2019:1–14. doi:10.1007/s00394-019-02036-y

373. Orlich MJ, Singh PN, Sabaté J, et al. Vegetarian dietary patterns and mortality in adventist health study 2. *JAMA Intern Med.* 2013. doi:10.1001/jamainternmed.2013.6473

374. Tuohy KM, Fava F, Viola R. The Way to a Man's Heart Is through His Gut Microbiota: Dietary Pro- and Prebiotics for the Management of Cardiovascular Risk. In: *Proceedings of the Nutrition Society.* Vol 73. Cambridge University Press; 2014:172–185. doi:10.1017/S0029665113003911

375. Schiattarella GG, Sannino A, Toscano E, et al. Gut microbe-generated metabolite trimethylamine-N-oxide as cardiovascular risk biomarker: A systematic review and dose-response meta-analysis. *Eur Heart J.* 2017;38(39):2948–2956. doi:10.1093/EURHEARTJ/EHX342

376. Heianza Y, Ma W, Manson JE, Rexrode KM, Qi L. Gut microbiota metabolites and risk of major adverse cardiovascular disease events and death: A systematic review and meta-analysis of prospective studies. *J Am Heart Assoc.* 2017;6(7). doi:10.1161/JAHA.116.004947

377. Dambrova M, Latkovskis G, Kuka J, et al. Diabetes is associated with higher trimethylamine n-oxide plasma

levels. *Exp Clin Endocrinol Diabetes.* 2016;124(4):251–256. doi:10.1055/S-0035-1569330

378. Tang WHW, Wang Z, Kennedy DJ, et al. Gut microbiota-dependent trimethylamine-N-oxide (TMAO) pathway contributes to both development of renal insufficiency and mortality risk in chronic kidney disease. *Circ Res.* 2014;116(3):448–455. doi:10.1161/CIRCRESAHA.116.305360

379. Zhu W, Gregory JC, Org E, et al. Gut microbial metabolite TMAO enhances platelet hyperreactivity and thrombosis risk. *Cell.* 2016;165(1):111–124. doi:10.1016/j.cell.2016.02.011

380. Bae S, Ulrich CM, Neuhouser ML, et al. Plasma choline metabolites and colorectal cancer risk in the Women's Health Initiative Observational Study. *Cancer Res.* 2014;74(24): 7442–7452. doi:10.1158/0008-5472.CAN-14-1835

381. Fu BC, Hullar MAJ, Randolph TW, et al. Associations of plasma trimethylamine N-oxide, choline, carnitine, and betaine with inflammatory and cardiometabolic risk biomarkers and the fecal microbiome in the multiethnic cohort adiposity phenotype study. *Am J Clin Nutr.* 2020;111(6):1226–1234. doi:10.1093/ajcn/nqaa015

382. Xu R, Wang QQ, Li L. A genome-wide systems analysis reveals strong link between colorectal cancer and trimethylamine N-oxide (TMAO), a gut microbial metabolite of dietary meat and fat. *BMC Genomics.* 2015;16(Suppl 7):S4. doi:10.1186/1471-2164-16-S7-S4

383. Papandreou C, Moré M, Bellamine A. Trimethylamine N-oxide in relation to cardiometabolic health: Cause or effect? *Nutrients.* 2020;12(5). doi:10.3390/NU12051330

384. Martínez-González MA, Sánchez-Tainta A, Corella D, et al. A Provegetarian Food Pattern and Reduction in Total Mortality in

the Prevención Con Dieta Mediterránea (PREDIMED) Study. In: *American Journal of Clinical Nutrition*. Vol 100. American Society for Nutrition; 2014. doi:10.3945/ajcn.113.071431

385. Velasquez MT, Ramezani A, Manal A, Raj DS. Trimethylamine N-oxide: The good, the bad and the unknown. *Toxins (Basel)*. 2016;8(11). doi:10.3390/toxins8110326

386. Guasch-Ferré M, Satija A, Blondin SA, et al. Meta-analysis of randomized controlled trials of red meat consumption in comparison with various comparison diets on cardiovascular risk factors. *Circulation*. 2019;139(15):1828–1845. doi:10.1161/ CIRCULATIONAHA.118.035225

387. Crimarco A, Springfield S, Petlura C, et al. A randomized crossover trial on the effect of plant-based compared with animal-based meat on trimethylamine N-oxide and cardiovascular disease risk factors in generally healthy adults: Study with appetizing plantfood: Meat eating alternative trial (SWAP-ME). *Am J Clin Nutr*. 2020. doi:10.1093/AJCN/NQAA203

388. ATSDR Polychlorinated Biphenyls (PCBs). Public Health Statement. https://wwwn.cdc.gov/TSP/PHS/PHS. aspx?phsid=139&toxid=26. Accessed February 4, 2022.

389. Food Safety: Persistent Organic Pollutants (POPs). https://www. who.int/news-room/questions-and-answers/item/food-safety- persistent-organic-pollutants-(pops). Accessed February 3, 2022.

390. Carrasco Cabrera L, Medina Pastor P. The 2019 European union report on pesticide residues in food. *EFSA J*. 2021;19(4). doi:10.2903/J.EFSA.2021.6491

391. Kumar M, Sarma DK, Shubham S, et al. Environmental endocrine-disrupting chemical exposure: Role in non- communicable diseases. *Front Public Heal*. 2020;8:553850. doi:10.3389/fpubh.2020.553850

392. Sandoval-Insausti H, Chiu YH, Lee DH, et al. Intake of fruits and vegetables by pesticide residue status in relation to cancer risk. *Environ Int.* 2021;156. doi:10.1016/j.envint.2021.106744

393. Matich EK, Laryea JA, Seely KA, Stahr S, Su LJ, Hsu PC. Association between pesticide exposure and colorectal cancer risk and incidence: A systematic review. *Ecotoxicol Environ Saf.* 2021;219. doi:10.1016/j.ecoenv.2021.112327

394. Cohn BA, Wolff MS, Cirillo PM, Scholtz RI. DDT and breast cancer in young women: New data on the significance of age at exposure. *Environ Health Perspect.* 2007;115(10):1406–1414. doi:10.1289/EHP.10260

395. Rude KM, Pusceddu MM, Keogh CE, et al. Developmental exposure to polychlorinated biphenyls (PCBs) in the maternal diet causes host-microbe defects in weanling offspring mice. *Environ Pollut.* 2019;253:708–721. doi:10.1016/J.ENVPOL.2019.07.066

396. Tsiaoussis J, Antoniou MN, Koliarakis I, et al. Effects of single and combined toxic exposures on the gut microbiome: Current knowledge and future directions. *Toxicol Lett.* 2019;312:72–97. doi:10.1016/j.toxlet.2019.04.014

397. Mesnage R, Teixeira M, Mandrioli D, et al. Shotgun metagenomics and metabolomics reveal glyphosate alters the gut microbiome of Sprague-Dawley rats by inhibiting the shikimate pathway. *bioRxiv.* December 2019:870105. doi:10.1101/870105

398. Barnett JA, Gibson DL. Separating the empirical wheat from the pseudoscientific chaff: A critical review of the literature surrounding glyphosate, dysbiosis and wheat-sensitivity. *Front Microbiol.* 2020;11. doi:10.3389/fmicb.2020.556729

399. Liang Y, Zhan J, Liu D, et al. Organophosphorus pesticide chlorpyrifos intake promotes obesity and insulin resistance through impacting gut and gut microbiota. *Microbiome.* 2019;7(1). doi:10.1186/s40168-019-0635-4

400. Tsiaoussis J, Antoniou MN, Koliarakis I, et al. Effects of single and combined toxic exposures on the gut microbiome: Current knowledge and future directions. 2019. doi:10.1016/j.toxlet.2019.04.014

401. Gao J, Ellis LBM, Wackett LP. The University of Minnesota Biocatalysis/Biodegradation database: Improving public access. *Nucleic Acids Res.* 2010;38(Database issue):D488. doi:10.1093/NAR/GKP771

402. Helmus DS, Thompson CL, Zelenskiy S, Tucker TC, Li L. Red meat-derived heterocyclic amines increase risk of colon cancer: A population-based case-control study. *Nutr Cancer.* 2013;65(8):1141–1150. doi:10.1080/01635581.2013.834945

403. Nogacka AM, Gómez-Martín M, Suárez A, González-Bernardo O, de los Reyes-Gavilán CG, González S. Xenobiotics formed during food processing: Their relation with the intestinal Microbiota and colorectal cancer. *Int J Mol Sci.* 2019;20(8):2051. doi:10.3390/ijms20082051

404. Humblot C, Murkovic M, Rigottier-Gois L, et al. β-glucuronidase in human intestinal microbiota is necessary for the colonic genotoxicity of the food-Borne carcinogen 2-amino-3-methylimidazo[4,5-f]quinoline in rats. *Carcinogenesis.* 2007;28(11):2419–2425. doi:10.1093/carcin/bgm170

405. Kassie F, Rabot S, Kundi M, Chabicovsky M, Qin HM, Knasmüller S. Intestinal microflora plays a crucial role in the genotoxicity of the cooked food mutagen

2-amino-3-methylimidazo [4,5-f]quinoline. *Carcinogenesis.* 2001;22(10):1721–1725. doi:10.1093/CARCIN/22.10.1721

406. Koppel N, Rekdal VM, Balskus EP. Chemical transformation of xenobiotics by the human gut microbiota. *Science.* 2017;356(6344):1246–1257. doi:10.1126/SCIENCE.AAG2770

407. Scalbert A, Brennan L, Manach C, et al. The food metabolome: A window over dietary exposure. *Am J Clin Nutr.* 2014;99(6): 1286–1308. doi:10.3945/AJCN.113.076133

408. Diener C, Qin S, Zhou Y, et al. Baseline gut metagenomic functional gene signature associated with variable weight loss responses following a healthy lifestyle intervention in humans. Ercolini D, ed. *mSystems.* September 2021. doi:10.1128/ MSYSTEMS.00964-21

409. Tang ZZ, Chen G, Hong Q, et al. Multi-omic analysis of the microbiome and metabolome in healthy subjects reveals microbiome-dependent relationships between diet and metabolites. *Front Genet.* 2019;10(MAY):454. doi:10.3389/ fgene.2019.00454

410. Posma JM, Garcia-Perez I, Frost G, et al. Nutriome– metabolome relationships provide insights into dietary intake and metabolism. *Nat Food 2020 17.* 2020;1(7):426–436. doi:10.1038/s43016-020-0093-y

411. Fryc J, Naumnik B. Thrombolome and its emerging role in chronic kidney diseases. *Toxins (Basel).* 2021;13(3). doi:10.3390/ toxins13030223

412. Floegel A, Stefan N, Yu Z, et al. Identification of serum metabolites associated with risk of type 2 diabetes using a targeted metabolomic approach. *Diabetes.* 2013;62(2):639–648. doi:10.2337/db12-0495

413. Lever M, George PM, Slow S, et al. Betaine and trimethylamine N-oxide as predictors of cardiovascular outcomes show different patterns in diabetes mellitus: An observational study. *PLoS One*. 2014;9(12). doi:10.1371/JOURNAL.PONE.0114969

414. Eyileten C, Jarosz-Popek J, Jakubik D, et al. Plasma trimethylamine-N-oxide is an independent predictor of long-term cardiovascular mortality in patients undergoing percutaneous coronary intervention for acute coronary syndrome. *Front Cardiovasc Med*. 2021;8:728724. doi:10.3389/fcvm.2021.728724

415. Randrianarisoa E, Lehn-Stefan A, Wang X, et al. Relationship of serum trimethylamine N-oxide (TMAO) levels with early atherosclerosis in humans. *Sci Rep*. 2016;6. doi:10.1038/srep26745

416. Ottiger M, Nickler M, Steuer C, et al. Trimethylamine N-oxide (TMAO) predicts fatal outcomes in community-acquired pneumonia patients without evident coronary artery disease. *Eur J Intern Med*. 2016;36:67–73. doi:10.1016/j.ejim.2016.08.017

417. Ottiger M, Nickler M, Steuer C, et al. Gut, microbiota-dependent trimethylamine N-oxide is associated with long-term all-cause mortality in patients with exacerbated chronic obstructive pulmonary disease. *Nutrition*. 2018;45:135–141.e1. doi:10.1016/j.nut.2017.07.001

418. Zhuang Z, Li N, Wang J, et al. GWAS-associated bacteria and their metabolites appear to be causally related to the development of inflammatory bowel disease. *Eur J Clin Nutr*. January 2022. doi:10.1038/S41430-022-01074-W

419. Oh TG, Kim SM, Caussy C, et al. A universal gut-microbiome-derived signature predicts cirrhosis. *Cell Metab*. 2020;32(5):878–888.e6. doi:10.1016/J.CMET.2020.06.005

420. Vineis P, Wild CP. Global cancer patterns: Causes and prevention. *Lancet.* 2014;383(9916):549–557. doi:10.1016/S0140-6736(13)62224-2

421. Pasquet P. Theories of human Evolutionary trends in meat eating and studies of primate intestinal tracts. https://www.researchgate.net/publication/267220566_Theories_of_Human_Evolutionary_Trends_in_Meat_Eating_and_Studies_of_Primate_Intestinal_Tracts. Accessed November 2, 2021.

422. Marcel Hladik C, Pasquet P. *The Human Adaptations to Meat Eating: A Reappraisal.* Vol 17. Springer Verlag; 2002. https://hal.archives-ouvertes.fr/hal-00545795. Accessed December 3, 2019.

423. Marcel Hladik C. Seasonal variations in food supply for wild primates. 1988:1–25. https://hal.archives-ouvertes.fr/hal-00578690. Accessed November 2, 2021.

424. Furness JB, Cottrell JJ, Bravo DM. Comparative gut physiology symposium: Comparative physiology of digestion. *J Anim Sci.* 2015;93(2):485–491. doi:10.2527/JAS.2014-8481

425. Hladik, C.M.. Diet and the evolution of feeding strategies among forest primates. In (R.S.O. Harding and G. Teleki, eds) *Omnivorous Primates: Gathering and Hunting in Human Evolution.* Columbia Univ Press New York. 1981:215–254.

426. Jabri A, Kumar A, Verghese E, et al. Meta-analysis of effect of vegetarian diet on ischemic heart disease and all-cause mortality. *Am J Prev Cardiol.* 2021;7:100182. doi:10.1016/J.AJPC.2021.100182

427. Le LT, Sabaté J. Beyond meatless, the health effects of vegan diets: Findings from the adventist cohorts. *Nutrients.* 2014;6(6):2131–2147. doi:10.3390/nu6062131

428. Roberts WC. The cause of atherosclerosis. *Nutr Clin Pract.* 2008;23(5):464–467. doi:10.1177/0884533608324586

429. Roberts WC. Twenty questions on atherosclerosis. *Proc (Bayl Univ Med Cent).* 2000;13(2):139. doi:10.1080/08998280.2000.11927657

430. Chatterjee IB. Evolution and the biosynthesis of ascorbic acid. *Science.* 1973;182(4118):1271–1272. doi:10.1126/SCIENCE.182.4118.1271

431. Hornung TC, Biesalski H-K. Glut-1 explains the evolutionary advantage of the loss of endogenous vitamin C synthesis: The electron transfer hypothesis. *Evol Med Public Heal.* 2019;2019(1):221–231. doi:10.1093/EMPH/EOZ024

432. Lachapelle MY, Drouin G. Inactivation dates of the human and guinea pig vitamin C genes. *Genetica.* 2011;139(2):199–207. doi:10.1007/S10709-010-9537-X

433. The "True" Human Diet. Scientific American Blog Network. https://blogs.scientificamerican.com/guest-blog/the-true-human-diet/. Accessed November 3, 2021.

434. Leach JD. Evolutionary perspective on dietary intake of fibre and colorectal cancer. *Eur J Clin Nutr 2007 611.* 2006;61(1):140–142. doi:10.1038/sj.ejcn.1602486

435. Ungar, PS. Evolution's Bite: A Story of Teeth, Diet, and Human Origins. https://books.google.se/books?id=vViYDwAAQBAJ&pg=PA199&lpg=PA199&dq=pythagoras+Oh,+how+wrong+it+is+for+flesh+to+be+made+from+flesh&source=bl&ots=0SIy4_fafc&sig=ACfU3U07mbNy6SjDUvlP4JcfFQMfsteIPw&hl=de&sa=X&ved=2ahUKEwi76LGeoPzzAhXjAxAIHWC9B_4Q6AF6BAgdEAM#v=onepage&q=pythagoras Oh%2C how wrong it is for flesh to be made from flesh&f=false. Accessed November 3, 2021.

436. Human Ancestors were nearly all vegetarians. *Sci Am.* 2012. https://blogs.scientificamerican.com/guest-blog/human-ancestors-were-nearly-all-vegetarians/.

437. Ungar PS, Sponheimer M. The Diets of Early Hominins. www.sciencemag.org. Accessed November 4, 2021.

438. Eaton S, Eaton S, Konner M. Review Paleolithic nutrition revisited: A twelve-year retrospective on its nature and implications. *Eur J Clin Nutr.* 1997;51(4):207–216. doi:10.1038/sj.ejcn.1600389

439. Hardy K, Buckley S, Collins MJ, et al. Neanderthal medics? Evidence for food, cooking, and medicinal plants entrapped in dental calculus. *Naturwissenschaften.* 2012;99(8):617–626. doi:10.1007/S00114-012-0942-0

440. Kaplan M. Neanderthals ate their greens. *Nature.* July 2012. doi:10.1038/NATURE.2012.11030

441. Kuhnlein H V., Turner NJ. Traditional plant foods of Canadian indigenous peoples. *Tradit Plant Foods Can Indig Peoples.* October 2020. doi:10.4324/9781003054689

442. Kuhnlein H V. Nutrient values in indigenous wild plant greens and roots used by the Nuxalk people of Bella Coola, British Columbia. *J Food Compos Anal.* 1990;3(1):38–46. doi:10.1016/0889-1575(90)90007-9

443. Tuohy K, Gougoulias C, Shen Q, Walton G, Fava F, Ramnani P. Studying the human gut microbiota in the trans-omics era: Focus on metagenomics and metabonomics. *Curr Pharm Des.* 2009;15(13):1415–1427. doi:10.2174/138161209788168182

444. Over-consumption of meat-based protein. *Keystone Dialogues.* http://www.wri.org/. Accessed March 13, 2022.

445. SchmidhuberJ, Shetty, P. (2005). The nutrition transition to 2030. Why developing countries are likely to bear the major burden. Food Economics - Acta Agriculturae Scandinavica, Section C. 2. 10.1080/16507540500534812

446. FAO. FAO's Animal Production and Health Division: Meat & Meat Products. https://www.fao.org/ag/againfo/themes/en/meat/home.html. Accessed March 13, 2022.

447. How Much Protein Do We Need? *New York Times.* https://www.nytimes.com/2017/07/28/well/eat/how-much-protein-do-we-need.html. Accessed November 4, 2021.

448. Can You Get Too Much Protein? *New York Times.* https://www.nytimes.com/2016/12/06/well/eat/can-you-get-too-much-protein.html?_r=0. Accessed November 4, 2021.

449. Agostoni C, Bresson J-L, Fairweather-Tait S, et al. EFSA Panel on Dietetic Products, Nutrition, and Allergies (NDA). Scientific opinion on dietary reference values for protein released for public consultation scientific opinion on dietary reference values for protein 1. *Nutr. EFSA J.* doi:10.2903/j.efsa.20NN.NNNN

450. Ernstoff A, Stylianou KS, Sahakian M, et al. Towards win–win policies for healthy and sustainable diets in Switzerland. *Nutrients.* 2020;12(9):1–24. doi:10.3390/nu12092745

451. Losasso C, Eckert EM, Mastrorilli E, et al. Assessing the influence of vegan, vegetarian and omnivore oriented westernized dietary styles on human gut microbiota: A cross sectional study. *Front Microbiol.* 2018;9:317. doi:10.3389/fmicb.2018.00317

452. Tomova A, Bukovsky I, Rembert E, et al. The effects of vegetarian and vegan diets on gut microbiota. *Front Nutr.* 2019;6:47. doi:10.3389/fnut.2019.00047

453. Wong M-W, Yi C-H, Liu T-T, et al. Impact of vegan diets on gut microbiota: An update on the clinical implications. *Tzu-Chi Med J.* 2018;30(4):200. doi:10.4103/tcmj.tcmj_21_18

454. Zimmer J, Lange B, Frick JS, et al. A vegan or vegetarian diet substantially alters the human colonic faecal microbiota. *Eur J Clin Nutr.* 2012;66(1):53–60. doi:10.1038/EJCN.2011.141

455. Hippe B, Zwielehner J, Liszt K, Lassl C, Unger F, Haslberger AG. Quantification of butyryl CoA:acetate CoA-transferase genes reveals different butyrate production capacity in individuals according to diet and age. *FEMS Microbiol Lett.* 2011;316(2):130–135. doi:10.1111/J.1574-6968.2010.02197.X

456. De Filippis F, Pellegrini N, Vannini L, et al. High-level adherence to a Mediterranean diet beneficially impacts the gut microbiota and associated metabolome. *Gut.* 2016;65(11): 1812–1821. doi:10.1136/gutjnl-2015-309957

457. Manor O, Zubair N, Conomos MP, et al. A multi-omic association study of trimethylamine N-oxide. *Cell Rep.* 2018;24(4):935–946. doi:10.1016/J.CELREP.2018.06.096

458. Cotillard A, Kennedy SP, Kong LC, et al. Dietary intervention impact on gut microbial gene richness. *Nature.* 2013;500(7464): 585–588. doi:10.1038/nature12480

459. Valeur J, Puaschitz NG, Midtvedt T, Berstad A. Oatmeal porridge: Impact on microflora-associated characteristics in healthy subjects. *Br J Nutr.* 2016;115(1):62–67. doi:10.1017/S0007114515004213

460. Keim NL, Martin RJ. Dietary Whole Grain-Microbiota Interactions: Insights into Mechanisms for Human Health. In: *Advances in Nutrition.* Vol 5. American Society for Nutrition; 2014:556–557. doi:10.3945/an.114.006536

461. Vitaglione P, Mennella I, Ferracane R, et al. Whole-grain wheat consumption reduces inflammation in a randomized controlled trial on overweight and obese subjects with unhealthy dietary and lifestyle behaviors: Role of polyphenols

bound to cereal dietary fiber. *Am J Clin Nutr.* 2015;101(2): 251–261. doi:10.3945/ajcn.114.088120

462. D'Argenio V, Precone V, Casaburi G, et al. An altered gut microbiome profile in a child affected by Crohn's disease normalized after nutritional therapy. *Am J Gastroenterol.* 2013;108(5):851–852. doi:10.1038/AJG.2013.46

463. Żyła E, Dziendzikowska K, Kamola D, et al. Anti-inflammatory activity of oat beta-glucans in a Crohn's disease model: Time- and molar mass-dependent effects. *Int J Mol Sci.* 2021;22(9). doi:10.3390/IJMS22094485

464. Chiba M, Nakane K, Tsuji T, et al. Relapse prevention in ulcerative colitis by plant-based diet through educational hospitalization: A single-group trial. *Perm J.* 2018;22. doi:10.7812/TPP/17-167

465. Chiba M, Nakane K, Komatsu M. Westernized diet is the most ubiquitous environmental factor in inflammatory bowel disease. *Perm J.* 2019;23:18–107. doi:10.7812/TPP/18-107

466. Ojo O, Feng QQ, Ojo OO, Wang XH. The role of dietary fibre in modulating gut microbiota dysbiosis in patients with type 2 diabetes: A systematic review and meta-analysis of randomised controlled trials. *Nutrients.* 2020;12(11):1–21. doi:10.3390/NU12113239

467. Zhao L, Zhang F, Ding X, et al. Gut bacteria selectively promoted by dietary fibers alleviate type 2 diabetes. *Science.* 2018;359(6380):1151–1156. doi:10.1126/science.aao5774

468. de Vries J, Miller PE, Verbeke K. Effects of cereal fiber on bowel function: A systematic review of intervention trials. *World J Gastroenterol.* 2015;21(29):8952–8963. doi:10.3748/wjg.v21.i29.8952

469. Jalanka J, Major G, Murray K, et al. The effect of psyllium husk on intestinal microbiota in constipated patients and healthy controls. *Int J Mol Sci Artic.* doi:10.3390/ijms20020433

470. Mars RAT, Yang Y, Ward T, et al. Longitudinal multi-omics reveals subset-specific mechanisms underlying irritable bowel syndrome. *Cell.* 2020;182(6):1460–1473.e17. doi:10.1016/J. CELL.2020.08.007/ATTACHMENT/9FF21193-1599-4A01-9A3B-3D67A0F71B92/MMC7.XLSX

471. Mancabelli L, Milani C, Lugli GA, et al. Unveiling the gut microbiota composition and functionality associated with constipation through metagenomic analyses. *Sci Rep.* 2017;7(1):9879. doi:10.1038/s41598-017-10663-w

472. Lewis SJ, Heaton KW. The metabolic consequences of slow colonic transit. *Am J Gastroenterol.* 1999;94(8):2010–2016. doi:10.1111/J.1572-0241.1999.01271.X

473. Maruti SS, Lampe JW, Potter JD, Ready A, White E. A prospective study of bowel motility and related factors on breast cancer risk. *Cancer Epidemiol Biomarkers Prev.* 2008;17(7):1746–1750. doi:10.1158/1055-9965.EPI-07-2850

474. Citronberg J, Kantor ED, Potter JD, White E. A prospective study of the effect of bowel movement frequency, constipation and laxative use on colorectal cancer risk. *Am J Gastroenterol.* 2014;109(10):1640–1649. doi:10.1038/ajg.2014.233

475. Stanaway JD, Afshin A, Gakidou E, et al. Global, regional, and national comparative risk assessment of 84 behavioural, environmental and occupational, and metabolic risks or clusters of risks for 195 countries and territories, 1990–2017: A systematic analysis for the Global Burden of Disease Study 2017. *Lancet.* 2018;392(10159):1923–1994. doi:10.1016/S0140-6736(18)32225-6

476. Clarys P, Deliens T, Huybrechts I, et al. Comparison of nutritional quality of the vegan, vegetarian, semi-vegetarian, pesco-vegetarian and omnivorous diet. *Nutrients.* 2014;6(3):1318–1332. doi:10.3390/NU6031318

477. Dinu M, Abbate R, Gensini GF, Casini A, Sofi F. Vegetarian, vegan diets and multiple health outcomes: A systematic review with meta-analysis of observational studies. *Crit Rev Food Sci Nutr.* 2017;57(17):3640–3649. doi:10.1080/10408398.2016.1138447

478. Holscher HD. Dietary fiber and prebiotics and the gastrointestinal microbiota. *Gut Microbes.* 2017;8(2):172–184. doi:10.1080/19490 976.2017.1290756

479. Holscher HD, Guetterman HM, Swanson KS, et al. Walnut consumption alters the gastrointestinal microbiota, microbially derived secondary bile acids, and health markers in healthy adults: A randomized controlled trial. *J Nutr.* 2018;148(6):861–867. doi:10.1093/jn/nxy004

480. Jacobs DR, Pereira MA, Kushi LH, Meyer KA. Fiber from whole grains, but not refined grains, is inversely associated with all-cause mortality in older women: The Iowa Women's Health Study. *J Am Coll Nutr.* 2000;19(3 Suppl):326S–330S. doi:10.1080/07315724.2000.10718968

481. Zinöcker MK, Lindseth IA. The western diet—microbiome-host interaction and its role in metabolic disease. *Nutrients.* 2018;10(3):365. doi:10.3390/nu10030365

482. Begley M, Hill C, Gahan CGM. Bile salt hydrolase activity in probiotics. *Appl Environ Microbiol.* 2006;72(3):1729–1738. doi:10.1128/AEM.72.3.1729-1738.2006

483. Pirman T, Ribeyre MC, Mosoni L, et al. Dietary pectin stimulates protein metabolism in the digestive tract. *Nutrition.* 2007;23(1):69–75. doi:10.1016/J.NUT.2006.09.001

484. Goodlad RA. Fiber can make your gut grow. *Nutrition*. 2007;23(5):434–435. doi:10.1016/J.NUT.2006.10.010

485. Tuohy KM, Conterno L, Gasperotti M, Viola R. Up-regulating the human intestinal microbiome using whole plant foods, polyphenols, and/or fiber. *J Agric Food Chem*. 2012;60(36): 8776–8782. doi:10.1021/JF2053959

486. Costabile A, Klinder A, Fava F, et al. Whole-grain wheat breakfast cereal has a prebiotic effect on the human gut microbiota: A double-blind, placebo-controlled, crossover study. *Br J Nutr*. 2008;99(1):110–120. doi:10.1017/ S0007114507793923

487. Vitaglione P, Napolitano A, Fogliano V. Cereal dietary fibre: A natural functional ingredient to deliver phenolic compounds into the gut. *Trends Food Sci Technol*. 2008;19(9):451–463. doi:10.1016/J.TIFS.2008.02.005

488. Armet AM, Deehan EC, Thöne J V, Hewko SJ, Walter J. The effect of isolated and synthetic dietary fibers on markers of metabolic diseases in human intervention studies: A systematic review. *Adv Nutr*. 2020;11(2):420–438. doi:10.1093/ADVANCES/NMZ074

489. Globally, one in five deaths are associated with poor diet. *ScienceDaily*. https://www.sciencedaily.com/ releases/2019/04/190403193702.htm. Accessed June 10, 2021.

490. Slavin JL. Position of the American Dietetic Association: Health implications of dietary fiber. *J Am Diet Assoc*. 2008;108(10):1716–1731. doi:10.1016/J.JADA.2008.08.007

491. Slavin J. Fiber and prebiotics: Mechanisms and health benefits. *Nutrients*. 2013;5(4):1417. doi:10.3390/NU5041417

492. Medicine i of. Dietary reference intakes for energy, carbohydrate, fiber, fat, fatty acids, cholesterol, protein, and amino acids. *Diet Ref Intakes Energy, Carbohydrate, Fiber, Fat, Fat Acids,*

Cholesterol, Protein, Amino Acids. September 2002:1–1331. doi:10.17226/10490

493. Tantamango YM, Knutsen SF, Beeson WL, Fraser G, Sabate J. Foods and food groups associated with the incidence of colorectal polyps: The Adventist Health Study. *Nutr Cancer.* 2011. doi:10.1080/01635581.2011.551988

494. Ornish D, Weidner G, Fair WR, et al. Intensive lifestyle changes may affect the progression of prostate cancer. *J Urol.* 2005;174(3):1065–1070. doi:10.1097/01.JU.0000169487. 49018.73

495. "In Defense of Food": Author Offers Advice for Health. NPR. https://www.npr.org/templates/story/story.php?storyId=177259 32&t=1636124032679. Accessed November 5, 2021.

496. Song M, Fung TT, Hu FB, et al. Association of animal and plant protein intake with all-cause and cause-specific mortality. *JAMA Intern Med.* 2016;176(10):1453–1463. doi:10.1001/ JAMAINTERNMED.2016.4182

497. Estruch R, Ros E, Salas-Salvadó J, et al. Primary prevention of cardiovascular disease with a Mediterranean diet. *N Engl J Med.* 2013;368(14):1279–1290. doi:10.1056/ NEJMoa1200303

498. Madigan M, Karhu E. The role of plant-based nutrition in cancer prevention. *J Unexplored Med Data.* 2018;3(11):9. doi:10.20517/2572-8180.2018.05

499. Blue Zones. Live Longer, Better. https://www.bluezones. com/#section-1. Accessed June 16, 2021.

500. Metchnikoff E. *The Prolongation of Life: Optimistic Studies by Elie Metchnikoff.*; 1908. https://www.gutenberg.org/ebooks/51521. Accessed March 21, 2022.

501. Markets and Markets. Probiotics Market Growth Analysis, Trends, and Forecasts to 2026. https://www. marketsandmarkets.com/Market-Reports/probiotic-market-advanced-technologies-and-global-market-69.html. Accessed November 16, 2021.

502. Hill C, Guarner F, Reid G, et al. The International Scientific Association for Probiotics and Prebiotics consensus statement on the scope and appropriate use of the term probiotic. *Nat Rev Gastroenterol Hepatol.* 2014;11(8):506–514. doi:10.1038/nrgastro.2014.66

503. Rather IA, Bajpai VK, Kumar S, Lim J, Paek WK, Park YH. Probiotics and atopic dermatitis: An overview. *Front Microbiol.* 2016;7(APR). doi:10.3389/FMICB.2016.00507

504. Meneghin F, Fabiano V, Mameli C, Zuccotti GV. Probiotics and atopic dermatitis in children. *Pharmaceuticals (Basel).* 2012;5(7):727–744. doi:10.3390/PH5070727

505. Foolad N, Armstrong AW. Prebiotics and probiotics: The prevention and reduction in severity of atopic dermatitis in children. *Benef Microbes.* 2014;5(2):151–160. doi:10.3920/BM2013.0034

506. Puebla-Barragan S, Reid G. Forty-five-year evolution of probiotic therapy. *Microb Cell.* 2019;6(4):184–196. doi:10.15698/mic2019.04.673

507. Athalye-Jape G, Deshpande G, Rao S, Patole S. Benefits of probiotics on enteral nutrition in preterm neonates: A systematic review. *Am J Clin Nutr.* 2014;100(6):1508–1519. doi:10.3945/AJCN.114.092551

508. Laitinen K, Kalliomäki M, Poussa T, Lagström H, Isolauri E. Evaluation of diet and growth in children with and without

atopic eczema: Follow-up study from birth to 4 years. *Br J Nutr.* 2005;94(4):565–574. doi:10.1079/BJN20051503

509. Kopp MV, Hennemuth I, Heinzmann A, Urbanek R. Randomized, double-blind, placebo-controlled trial of probiotics for primary prevention: No clinical effects of lactobacillus GG supplementation. *Pediatrics.* 2008;121(4). doi:10.1542/PEDS.2007-1492

510. Cai J, Zhao C, Du Y, Zhang Y, Zhao M, Zhao Q. Comparative efficacy and tolerability of probiotics forantibiotic-associated diarrhea: Systematic review with networkmeta-analysis. *United Eur Gastroenterol J.* 2018;6(2):169. doi:10.1177/2050640617736987

511. Derwa Y, Gracie DJ, Hamlin PJ, Ford AC. Systematic review with meta-analysis: The efficacy of probiotics in inflammatory bowel disease. *Aliment Pharmacol Ther.* 2017. doi:10.1111/apt.14203

512. Tasson L, Canova C, Vettorato MG, Savarino E, Zanotti R. Influence of diet on the course of inflammatory bowel disease. *Dig Dis Sci.* 2017;62(8):2087–2094. doi:10.1007/s10620-017-4620-0

513. Ford AC, Harris LA, Lacy BE, Quigley EMM, Moayyedi P. Systematic review with meta-analysis: The efficacy of prebiotics, probiotics, synbiotics and antibiotics in irritable bowel syndrome. *Aliment Pharmacol Ther.* 2018;48(10): 1044–1060. doi:10.1111/APT.15001

514. Zhang Y, Li L, Guo C, et al. Effects of probiotic type, dose and treatment duration on irritable bowel syndrome diagnosed by Rome III criteria: A meta-analysis. *BMC Gastroenterol.* 2016;16(1):62. doi:10.1186/s12876-016-0470-z

515. Silk DBA, Davis A, Vulevic J, Tzortzis G, Gibson GR. Clinical trial: The effects of a trans-galactooligosaccharide prebiotic

on faecal microbiota and symptoms in irritable bowel syndrome. *Aliment Pharmacol Ther.* 2009;29(5):508–518. doi:10.1111/J.1365-2036.2008.03911.X

516. Dreher ML. Whole fruits and fruit fiber emerging health effects. *Nutrients.* 2018;10(12). doi:10.3390/NU10121833

517. Costello EK, Stagaman K, Dethlefsen L, Bohannan BJM, Relman DA. The application of ecological theory toward an understanding of the human microbiome. *Science.* 2012;336(6086):1255–1262. doi:10.1126/SCIENCE.1224203

518. Maldonado-Gómez MX, Martínez I, Bottacini F, et al. Stable engraftment of *Bifidobacterium longum* AH1206 in the human gut depends on individualized features of the resident microbiome. *Cell Host Microbe.* 2016;20(4):515–526. doi:10.1016/J.CHOM.2016.09.001

519. Lieberman TD. Seven billion microcosms: Evolution within human microbiomes. *mSystems.* 2018;3(2). doi:10.1128/MSYSTEMS.00171-17

520. *EFSA Remit & Role: With Focus on Scientific Substantiation of Health Claims Made on Foods EFSA Meeting with IPA Europe.*

521. Macedo F, Fredua-Agyeman M. Evaluation of commercial probiotic products. *Br J Pharm.* 2016;1(1). doi:10.5920/BJPHARM.2016.11

522. Drago L, Rodighiero V, Celeste T, Rovetto L, de Vecchi E. Microbiological evaluation of commercial probiotic products available in the USA in 2009. *J Chemother.* 2010;22(6):373–377. doi:10.1179/JOC.2010.22.6.373

523. Suez J, Zmora N, Zilberman-Schapira G, et al. Post-antibiotic gut mucosal microbiome reconstitution is impaired

by probiotics and improved by autologous FMT. *Cell.* 2018;174(6):1406–1423.e16. doi:10.1016/j.cell.2018.08.047

524. Redman MG, Ward EJ, Phillips RS. The efficacy and safety of probiotics in people with cancer: A systematic review. *Ann Oncol.* 2014;25(10):1919–1929. doi:10.1093/ANNONC/MDU106

525. Zawistowska-Rojek A, Tyski S. Are probiotic really safe for humans? *Polish J Microbiol.* 2018;67(3):251–258. doi:10.21307/pjm-2018-044

526. Bafeta A, Koh M, Riveros C, Ravaud P. Harms reporting in randomized controlled trials of interventions aimed at modifying microbiota: A systematic review. *Ann Intern Med.* 2018;169(4):240–247. doi:10.7326/M18-0343

527. Bhide A, Datar S. Fecal microbiota transplants (FMT): Case histories of significant medical advances. Harvard Business School Working Paper, No. 21–132, June 2021.

528. Kleger A, Schnell J, Essig A, et al. Fecal transplant in refractory *Clostridium difficile* colitis. *Dtsch Arztebl Int.* 2013;110(7): 108–115. doi:10.3238/arztebl.2013.0108

529. Fecal enema as an adjunct in the treatment of pseudomembranous enterocolitis. PubMed. https://pubmed.ncbi.nlm.nih.gov/13592638/. Accessed November 18, 2021.

530. Kassam Z, Lee CH, Yuan Y, Hunt RH. Fecal microbiota transplantation for *Clostridium difficile* infection: Systematic review and meta-analysis. *Am J Gastroenterol.* 2013;108(4): 500–508. doi:10.1038/AJG.2013.59

531. Holvoet T, Joossens M, Vázquez-Castellanos JF, et al. Fecal microbiota transplantation reduces symptoms in some patients with irritable bowel syndrome with predominant abdominal bloating: Short- and long-term results from a placebo-controlled randomized trial. *Gastroenterology.* 2021;160(1): 145–157.e8. doi:10.1053/J.GASTRO.2020.07.013

532. Costello SP, Soo W, Bryant RV, Jairath V, Hart AL, Andrews JM. Systematic review with meta-analysis: Faecal microbiota transplantation for the induction of remission for active ulcerative colitis. *Aliment Pharmacol Ther*. 2017;46(3):213–224. doi:10.1111/apt.14173

533. Baruch EN, Youngster I, Ben-Betzalel G, et al. Fecal microbiota transplant promotes response in immunotherapy-refractory melanoma patients. *Science*. 2021;371(6529):602–609. doi:10.1126/SCIENCE.ABB5920/SUPPL_FILE/PAP.PDF

534. Kang DW, Adams JB, Gregory AC, et al. Microbiota transfer therapy alters gut ecosystem and improves gastrointestinal and autism symptoms: An open-label study. *Microbiome*. 2017. doi:10.1186/s40168-016-0225-7

535. Can patients' gut microbes help fight cancer? *Science* AAAS. https://www.science.org/content/article/can-patients-gut-microbes-help-fight-cancer. Accessed November 19, 2021.

536. Ridaura VK, Faith JJ, Rey FE, et al. Gut microbiota from twins discordant for obesity modulate metabolism in mice. *Science*. 2013;341(6150):1241214. doi:10.1126/science.1241214

537. Ott SJ, Waetzig GH, Rehman A, et al. Efficacy of sterile fecal filtrate transfer for treating patients with *Clostridium difficile* infection. *Gastroenterology*. 2017;152(4):799–811.e7. doi:10.1053/j.gastro.2016.11.010

538. DeFilipp Z, Bloom PP, Torres Soto M, et al. Drug-resistant *E. coli* bacteremia transmitted by fecal microbiota transplant. *N Engl J Med*. 2019;381(21):2043–2050. doi:10.1056/NEJMOA1910437

539. Zellmer C, Sater MRA, Huntley MH, Osman M, Olesen SW, Ramakrishna B. Shiga toxin-producing *Escherichia coli* transmission via fecal microbiota transplant. *Clin Infect Dis*. 2021;72(11):E876–E880. doi:10.1093/cid/ciaa1486

540. Could Food Be a New Medicine to Fight Heart Disease? Cleveland HeartLab, Inc. https://www.clevelandheartlab.com/blog/horizons-could-food-be-a-new-medicine-to-fight-heart-disease/. Accessed November 19, 2021.

541. Wan YJY, Jena PK. Precision dietary supplementation based on personal gut microbiota. *Nat Rev Gastroenterol Hepatol.* 2019;16(4):204–206. doi:10.1038/S41575-019-0108-Z

542. Berry SE, Valdes AM, Drew DA, et al. Human postprandial responses to food and potential for precision nutrition. *Nat Med.* 2020;26(6):964–973. doi:10.1038/S41591-020-0934-0

543. Søndertoft NB, Vogt JK, Arumugam M, et al. The intestinal microbiome is a co-determinant of the postprandial plasma glucose response. *PLoS One.* 2020;15(9 September). doi:10.1371/journal.pone.0238648

544. Jiao F, Guo R, Beckmann JS, et al. Great future or greedy venture: Precision medicine needs philosophy. *Heal Sci Reports.* 2021;4(3). doi:10.1002/HSR2.376

545. Zeevi D, Korem T, Zmora N, et al. Personalized nutrition by prediction of glycemic responses. *Cell.* 2015;163(5):1079–1094. doi:10.1016/J.CELL.2015.11.001/ATTACHMENT/B1291C30-58CE-4EC8-8292-FC85E2C3DBC3/MMC2.PDF

546. Wolever TMS. Personalized nutrition by prediction of glycaemic responses: Fact or fantasy? *Eur J Clin Nutr.* 2016;70(4):411–413. doi:10.1038/EJCN.2016.31

547. McMillan-Price J, Petocz P, Atkinson F, et al. Comparison of 4 diets of varying glycemic load on weight loss and cardiovascular risk reduction in overweight and obese young adults: A randomized controlled trial. *Arch Intern Med.* 2006;166(14):1466–1475. doi:10.1001/ARCHINTE.166.14.1466

548. Chatelan A, Bochud M, Frohlich KL. Precision nutrition: Hype or hope for public health interventions to reduce obesity? *Int J Epidemiol.* 2019;48(2):332–342. doi:10.1093/IJE/DYY274

549. 16S Ribosomal RNA: An overview. *ScienceDirect Topics.* https://www.sciencedirect.com/topics/neuroscience/16s-ribosomal-rna. Accessed April 3, 2022.

550. Yarza P, Yilmaz P, Pruesse E, et al. Uniting the classification of cultured and uncultured bacteria and archaea using 16S rRNA gene sequences. *Nat Rev Microbiol.* 2014;12(9):635–645. doi:10.1038/nrmicro3330

551. Classification: Medical Microbiology. NCBI Bookshelf. https://www.ncbi.nlm.nih.gov/books/NBK8406/. Accessed April 3, 2022.

552. Arumugam M, Raes J, Pelletier E, et al. Enterotypes of the human gut microbiome. *Nature.* 2011;473(7346):174–180. doi:10.1038/nature09944

553. Dietary Reference Intakes Proposed Definition of Dietary Fiber. NCBI Bookshelf. https://www.ncbi.nlm.nih.gov/books/NBK223586/. Accessed April 3, 2022.

554. Kennedy EA, King KY, Baldridge MT. Mouse microbiota models: Comparing germ-free mice and antibiotics treatment as tools for modifying gut bacteria. *Front Physiol.* 2018;9(Oct):1534. doi:10.3389/FPHYS.2018.01534/BIBTEX

555. Postler TS, Ghosh S. Cell Metabolism Review Understanding the Holobiont: How Microbial Metabolites Affect Human Health and Shape the Immune System. 2017. doi:10.1016/j.cmet.2017.05.008

556. Clevers H. The intestinal crypt, a prototype stem cell compartment. *Cell.* 2013;154(2):274–284. doi:10.1016/J.CELL.2013.07.004

557. Empl MT, Kammeyer P, Ulrich R, et al. The influence of chronic L-carnitine supplementation on the formation of preneoplastic and atherosclerotic lesions in the colon and aorta of male F344 rats. *Arch Toxicol.* 2015;89(11):2079. doi:10.1007/S00204-014-1341-4

558. Klassen A, Faccio AT, Canuto GAB, et al. Metabolomics: Definitions and significance in systems biology. *Adv Exp Med Biol.* 2017;965:3–17. doi:10.1007/978-3-319-47656-8_1

559. Nitroso Compounds: An Overview. *ScienceDirect Topics.* https://www.sciencedirect.com/topics/earth-and-planetary-sciences/nitroso-compounds. Accessed April 4, 2022.

560. Ferguson LR. Natural and human-made mutagens and carcinogens in the human diet. *Toxicology.* 2002;181-182:79–82. doi:10.1016/S0300-483X(02)00258-5

561. Noncommunicable diseases. https://www.who.int/news-room/fact-sheets/detail/noncommunicable-diseases. Accessed June 19, 2021.

562. Operational Taxonomic Unit: An Overview. *ScienceDirect Topics.* https://www.sciencedirect.com/topics/medicine-and-dentistry/operational-taxonomic-unit. Accessed April 3, 2022.

563. Velasquez MT, Ramezani A, Manal A, Raj DS. Trimethylamine N-oxide: The good, the bad and the unknown. *Toxins (Basel).* 2016;8(11). doi:10.3390/TOXINS8110326

564. Xenobiotic: An Overview. *ScienceDirect Topics.* https://www.sciencedirect.com/topics/immunology-and-microbiology/xenobiotic. Accessed April 3, 2022.

565. Zhu Y, Bo Y, Liu Y. Dietary total fat, fatty acids intake, and risk of cardiovascular disease: A dose-response meta-analysis of cohort studies. *Lipids Health Dis.* 2019;18(1):1–14. doi:10.1186/s12944-019-1035-2

INDEX

217

Printed in the United States
by Baker & Taylor Publisher Services